Metal complexes in organic chemistry

Cambridge Texts in Chemistry and Biochemistry

GENERAL EDITORS

D. T. Elmore
Professor of Biochemistry
The Queen's University of Belfast

J. Lewis
Professor of Inorganic Chemistry
University of Cambridge

K. Schofield
Professor of Organic Chemistry
University of Exeter

J. M. Thomas
Professor of Physical Chemistry
University of Cambridge

Metal complexes in
organic chemistry

R. P. HOUGHTON

Senior Lecturer in Organic Chemistry
University College, Cardiff

CAMBRIDGE UNIVERSITY PRESS
Cambridge
London · New York · Melbourne

Published by the Syndics of the Cambridge University Press
The Pitt Building, Trumpington Street, Cambridge CB2 1RP
Bentley House, 200 Euston Road, London NW1 2DB
32 East 57th Street, New York, NY 10022, USA
296 Beaconsfield Parade, Middle Park, Melbourne 3206, Australia

First published 1979

⊛ Typeset by H Charlesworth & Co Ltd, Huddersfield
Printed in Great Britain at the University Press, Cambridge

ISBN 0 521 21992 2 hard covers
ISBN 0 521 29331 6 paperback

78-51685

Contents

Preface

There are many ways in which metal complexes are involved in organic chemistry. For example, they may be used in the purification of organic compounds, in the resolution of racemic mixtures, and for simplifying nuclear magnetic resonance spectra. In reactions they may be used as reactants or catalysts, they may be generated and function as reactive species, or they may be formed as products. In view of this wide involvement it is surprising that at the present time there does not appear to be a book devoted to those aspects of the chemistry of metal complexes that make this type of compound so important to the modern organic chemist. While the various physical and inorganic aspects of their chemistry are adequately covered by a wide range of books, and while specific applications of certain types of metal complexes to organic chemistry have been reviewed (in some cases, in book form), there is not a general survey of the various ways in which metal complexes are involved in organic chemistry. The aim of the author has been to provide one.

Because of the breadth of the subject it is clear that every aspect could not be covered in a book of this size; one notable omission (with the exception of the Ziegler-Natta polymerisation of alkenes) is the metal-catalysed polymerisation of unsaturated organic compounds. The author's policy, therefore, has been to emphasise principles rather than facts and, generally speaking, detailed factual information is presented only in order to illustrate principles and not for its own sake.

The book is intended primarily for advanced undergraduates, graduates at the start of a research career in organic chemistry, and practising organic chemists who wish to learn more about the chemistry of metal complexes. It is hoped, however, that the book will be of use to all chemists who are concerned with the interaction of metals with organic ligands.

The author is deeply indebted to Professor L. N. Owen of Imperial College, University of London, for (amongst other things) first arousing his interest in the applications of metal complexes to organic chemistry. He is most grateful to Dr P. G. Blake and Dr D. W. Clack for many valuable discussions, and to those other members of the Department of Chemistry, University College,

Cardiff, whose critical reading of the manuscript helped to enhance its clarity; Dr M. J. E. Hewlins read the entire manuscript, while Professor R. D. Gillard and Dr D. A. Wilson read chapters 1 and 3 respectively. Finally, he would like to thank Professor K. Schofield of the University of Exeter, who as editor gave valuable advice and counsel.

Department of Chemistry   R. P. Houghton
University College, Cardiff
1978

Abbreviations

Except where otherwise indicated in the text, the following abbreviations are used as indicated below.

acac	acetylacetonate
COD	cyclo-octa-1,5-diene
dipyr	2,2'-dipyridyl
DMSO	dimethyl sulphoxide
dpma	o-phenylenebis(dimethyl)arsine
E^+	electrophile
EDTA	N,N,N',N'-ethylenediaminetetra-acetate
en	ethylenediamine
gly	glycinate
HMPTA	hexamethylphosphoramide
L	ligand
M, M^{n+}, M:	metal
NBD	norbornadiene
NuH, Nu$^-$, Nu:	nucleophile
pd	1,3-propylenediamine
PTDA	N,N,N',N'-propylenediaminetetra-acetate
phen	1,10-phenanthroline
py	pyridine
TMEDA	N,N,N',N'-tetramethylethylenediamine
trien	triethylenetetra-amine
X	halogen

1 General principles

1.1. Definitions

Although it is difficult to devise a definition which is both precise and informative, it is generally accepted that a *metal complex* is a chemical species which contains a metal atom or ion bonded to a greater number of ions or molecules than would be expected from simple valency considerations. The ions or molecules that are bonded or co-ordinated with the metal are termed *ligands*, and depending upon the charges carried by the metal and its ligands, a metal complex may be neutral, anionic, or cationic. Ionic metal complexes are always associated, of course, with the appropriate number of gegen ions. The actual atom through which a ligand is bonded to a metal is termed the *ligand atom*.

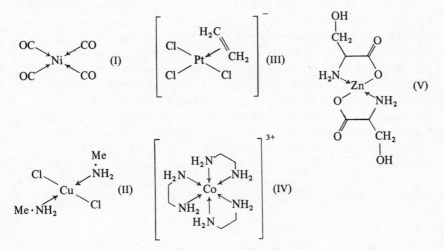

Some examples of metal complexes are given in structures (I)–(V), and as in almost all of the structures in this book the bonding between a metal and its ligands is represented as being covalent (M—L) for an anionic ligand, and co-ordinative or dative $(M \leftarrow L)$ for a neutral ligand. Although this type of representation is misleading in that it artificially differentiates between the metal–ligand bonds associated with the two types of ligand, it has the advantage that one can rapidly deduce the charge that would be left on the metal if all the ligands were removed with their ligand atoms in their closed-shell or 'inert gas' configuration, e.g. as Cl^-, H_2N^-, H_3N, or H_2O. This formal charge, or *oxidation number* of the metal, has proved to be of use in classifying the various types of metal complexes, for example with respect to stereochemistry, and is included in the name of the complex in the form of a Roman numeral placed in parentheses after the name of the metal. Structures (I) and (II), for example, are those of tetracarbonylnickel(0) and *trans*-dichlorobis(methylamine)copper(II) respectively.

In structures (IV) and (V) the ligands are bonded to the metal through two or more different ligand atoms and thus form parts of heterocyclic rings in which the metal is one of the members. The heterocyclic rings formed in this way are termed *chelate* rings, after the Greek word χηλή for a lobster's claw. A metal complex which contains one or more chelate rings is called a *metal chelate*, and the ligands involved are termed *chelating ligands*. The number of potential metal-bonding sites in a ligand is indicated by use of the terms *monodentate* (or *unidentate*), *bidentate*, *tridentate*, etc., but it should be noted that the stereochemistry of some ligands does not allow all the binding sites to be simultaneously bonded to the same metal. Thus, although serine is potentially tridentate (NH_2, OH, CO_2^-), in the chelate (V) it functions only as a bidentate ligand.

The *co-ordination number* of a metal in a complex is defined as the total number of ligand atoms bonded to the metal, e.g. the co-ordination numbers of nickel and cobalt in the complexes (I) and (IV) are four and six respectively. Although four and six are in fact the commonest, co-ordination numbers range from two to twelve and are determined by a number of factors, notably the size of the metal. Thus the maximum co-ordination number exhibited by the smaller metals lithium and beryllium is four, but this rises to six with transition metals in the first series and to eight with those in the third series. Any complex in which the co-ordination number of the metal can be increased by the addition of a further ligand is said to be *co-ordinatively unsaturated*. An example of such a complex is tetracarbonyl-iron(0) (VI), which is produced together with the corresponding penta-carbonyl (VII; L = CO) by the thermal decomposition of nonacarbonyldi-iron(0):

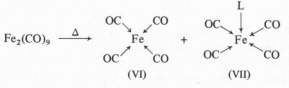

The normal co-ordination number of iron(0) is five, and accordingly the species (VI) reacts readily with many unidentate ligands (L), e.g. carbon monoxide or ammonia, to give pentaco-ordinate complexes of the type (VII).

With some types of co-ordinatively unsaturated complexes the co-ordination requirements of the metal are met by the complex existing as a dimer, trimer, or some other oligomeric form in which there are *bridging ligands*, i.e. ligands that are simultaneously bonded to more than one metal. Typical examples of bridging ligands are halide and alkoxide anions (see (VIII) and (IX) respectively).

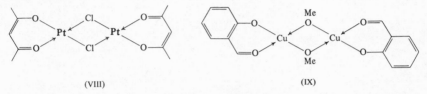

(VIII)   (IX)

The concepts of the co-ordination number of a metal and the denticity of a ligand were devised before the existence of *pi-* or *hapto-* (π or η) complexes was recognised. These are complexes in which the metal can be regarded as being bonded to two or more contiguous atoms of an unsaturated ligand by means of the π-electron system of the latter, e.g. tricarbonyl(η-allyl)cobalt(I) (X) and bis(η-benzene)chromium(0) (XI).

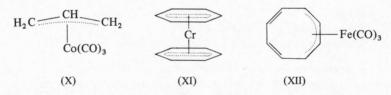

(X)   (XI)   (XII)

With most of these complexes there is usually little point in assigning a value to either the co-ordination number of the metal or the denticity of the ligand. However, if one wishes to indicate which particular atoms in an unsaturated ligand are considered to be bonded to the metal, one can insert either the total number of bonded atoms, or the positional numbers in a collective form, into the name of the complex. Thus (XII) can be called either tricarbonyl(η^4-cyclo-octatetraene)iron(0) or tricarbonyl(1-4-η-cyclo-octatetraene)iron(0).

1.2.   σ-Donor ligands

With the exception of those unsaturated ligands that use their π-bonded electrons to form π- or η-complexes, all ligands in the unco-ordinated state possess a pair of electrons in an orbital that is not involved in the bonding, but which can overlap with an empty orbital on a metal to form a metal-ligand σ-bond, i.e. the ligand acts as a σ-donor. Apart from the lanthanides and actinides where f-orbitals are often involved, the empty metal orbital can usually be regarded as a hybrid of suitable s- and p-, or s-, p- and d-orbitals. The four metal–ligand σ-bonds in the complex (XIII), for example, can be visualised as being formed by overlap of four empty $2s2p^3$ -hybrid orbitals on Be^{2+} with a filled 3p-atomic orbital on each of two chlorine anions and an oxygen lone pair $(2s2p^3)$ on each of two diethyl ether molecules.

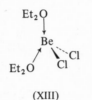

(XIII)

As illustrated by the ligands in the complex (XIII), the ligand atom of a σ-donor can be either neutral and bear a lone pair of electrons (H_2O:, H_3N:, Ph_3P:) or bear a formal negative charge (Cl^-, MeO^-, PhS^-). Many neutral and anionic species can act as σ-donors, but before considering specific examples it should be noted that σ-donors can be divided into three distinct types, i.e. those which can act as:

(a) σ-donors only,
(b) σ-donors and π-donors,
(c) σ-donors and π-acceptors.

Ligands of types (b) and (c) possess additional orbitals which can be used to form π-bonds with the metal. Those of type (b) donate electrons into an empty metal orbital while those of type (c) accept electrons from a filled metal orbital. The ligand H_2N^-, for example, can act as a σ-donor and as a π-donor, for it has two filled orbitals (sp^2 and p) both of which can participate in bonding if two suitable empty orbitals are available on the metal. One pair of metal–ligand orbitals gives rise to a σ-bond, the other pair to a π-bond, as shown in structure (XIV). In contrast, ligands of type (a) such as H^- and NH_3 lack the orbitals necessary for π-bond formation, and they therefore become attached to a metal only by a σ-bond.

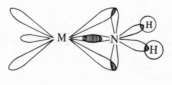

(XIV)

Although many different types of species can act as σ-donors the majority
of them have carbon, nitrogen, or oxygen as their ligand atom. The minority
have sulphur, phosphorus or, in a very small number of cases, an element of
comparatively low electronegativity such as silicon, tin, arsenic, or antimony.
The monatomic halide and hydride anions are also σ-donors, of course, and
the following is roughly the order in decreasing frequency in which the more
important ligand atoms are encountered in the chemistry of metal complexes:
halogens (mainly Cl or Br), O, N, C, S, P, H. Some of the common types of
unidentate σ-donors are listed in table 1.1. In every case the ligand atom is
identified by the presence of a negative charge or lone pair of electrons.

Bidentate ligands can σ-bond with a metal through two distinct ligand
atoms. If the stereochemistry of the ligand allows both these atoms to bond
simultaneously to the same metal then co-ordination results in a metal
chelate and the ligand is said to be a chelating ligand. Some typical bidentate
chelating ligands are given in table 1.2 and it can be seen that their donor
groups and associated ligand atoms are basically the same as those which are
present in the unidentate ligands listed in table 1.1. Some of the groups which
appear in table 1.1, however, do not appear in table 1.2, e.g. —C≡N and
—$\overset{+}{N}$≡$\bar{C}$. This is because most chelate rings are either 5- or 6-membered, with
7- and higher membered rings occurring very infrequently on account of their
comparatively low stability. As a result one observes that chelation never
involves those ligand atoms which bond to metals by means of an sp-hybrid
orbital, for it is geometrically impossible for the linear arrangement of the four
atoms in a co-ordinated nitrile (C—C≡N:→M), for example, to be incorporated
into a 5- or 6-membered ring.

For a ligand to form a 5- or 6-membered chelate ring the two ligand atoms
must be separated by two and three atoms respectively. These other atoms are
usually – but certainly not necessarily – carbon atoms in either an sp^2 or sp^3
state of hybridisation, as shown by all of the structures in table 1.2. It should
be noted that with all the anionic ligands in the table in which only one of the
ligands bears a negative charge, this charge cannot be delocalised by resonance
on to the other ligand atom, even in those ligands where the two ligand atoms
are separated by three sp^2-hybridised carbons. There are, however, a number
of important bidentate chelating anions in which delocalisation of the charge
can occur as shown below.

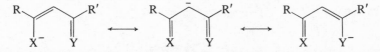

Included in this category are the anions of pentan-2,4-dione (acetylacetone,
XV), ethyl acetoacetate (XVI) and salicylaldehyde (XVII)), and the various
analogues obtained by replacing one or both of the ligand oxygen atoms in
these anions by either S or NR, e.g. (XVIII).

Table 1.1. *Some common types of unidentate ligands*

Ligand atom		Types of ligand
Carbon	$^-\text{C}\diagup_{\diagdown}$	Alkyl carbanions
	$_{\diagup}^-\text{C}=\text{C}_{\diagdown}^{\diagup}$	Alkenyl carbanions
	$^-\text{C}{\equiv}\text{C}-$	Alkynyl carbanions
	$^-\text{C}{\equiv}\text{N}$	Cyanide anion
	$^-\text{C}{\equiv}\overset{+}{\text{N}}-$	Isocyanides
	$^-\text{C}{\equiv}\overset{\cdot}{\text{O}}$	Carbon monoxide
Nitrogen	$:\text{N}\diagup_{\diagdown}$	Amines
	$^-\text{N}^{\diagup}_{\diagdown}$	Amide anions
	$\overset{\mid}{:\text{N}}{=}\text{C}^{\diagup}_{\diagdown}$	Imines, oximes, azines, hydrazones, unsaturated nitrogen heterocycles
	$^-\text{N}{=}\text{C}{=}\text{S}$	Isothiocyanate
	$:\text{N}{\equiv}\text{C}-$	Nitriles
	$:\text{N}{\equiv}\text{N}$	Dinitrogen
Oxygen	$:\overset{\cdot\cdot}{\text{O}}{\diagup}_{\diagdown}$	Water, alcohols, phenols, ethers
	$^-\text{O}-$	Hydroxide, alkoxide, and phenoxide anions
	$^-\text{O}_2\text{C}-$	Carboxylate anions
	$:\text{O}{=}\text{C}^{\diagup}_{\diagdown}$	Aldehydes, ketones, carboxylic acid derivatives
Sulphur	$^-\text{S}-$	Hydrosulphide, thiolate and thiophenolate anions
	$\text{S}{=}\text{C}^{\diagup}_{\diagdown}$	Thiocarbonyl compounds
Halogen	^-X	Halide anions
	$:\text{X}-$	Alkyl and aryl halides
	$:\text{X}\cdot\text{CO}-$	Acyl and aroyl halides
Phosphorus	$:\text{P}\diagup_{\diagdown}$	Phosphines
	$:\text{P}\overset{\text{O}-}{\underset{\text{O}-}{-\text{O}-}}$	Phosphites
Hydrogen	^-H	Hydride anion

Table 1.2. *Some examples of bidentate chelating ligands*

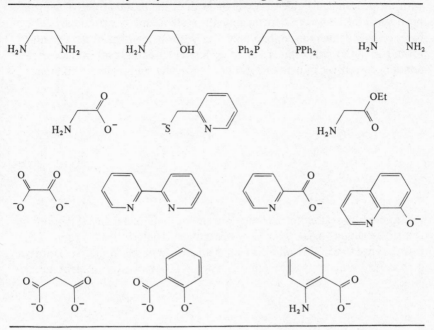

One general feature of these particular ligands is that with metal cations having a co-ordination number twice as large as the oxidation number, they form neutral complexes that are appreciably soluble in organic solvents. The anion of pentan-2,4-dione (acac) is particularly versatile in this respect and forms a very wide range of stable, soluble chelates of the types M(acac)$_2$ (M = Be, Cu, Zn), M(acac)$_3$ (M = Al, V, Fe), and M(acac)$_4$ (M = Hf, Th, U).

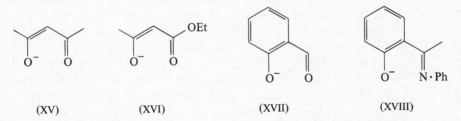

These chelates can be represented, of course, by resonance structures which reflect the resonance stabilisation of the parent ligand:

Another feature is that the protonated ligands are mixtures of tautomers, with the proton being able to reside on both of the ligand atoms. The proportions of the tautomers depend upon the system, and thus although Schiff bases of salicylaldehyde contain only a very low percentage of the enamine form, e.g. (XIX), this form is almost exclusively predominant in the corresponding derivatives of pentan-2,4-dione, e.g. (XX), and related β-diketones.

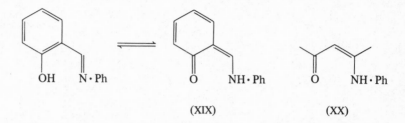

(XIX)   (XX)

Structurally related to the bidentate anions described above is the dipyrromethene anion (XXI), whose negative charge is delocalised over ten of the constituent atoms. This anion, which is invariably in the form of substituted derivatives, forms neutral chelates of type (XXII) with most bivalent metal ions.

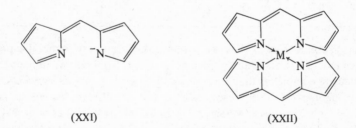

(XXI)   (XXII)

Delocalisation of charge over both ligand atoms (see XXIII*a* and *b*) also occurs in all the bidentate ligands that are known to form 4-membered ring chelates, although in fact such ligands are few in number. Examples are the dithiocarboxylate, xanthate, and dithiocarbamate anions (XXIII; Y = R, OR, and NR_2 respectively). With dithiocarbamate, and to a much smaller extent with xanthate, the ligand is also stabilised by a contribution from the resonance structure (XXIIIc), which is in accord with its π-donor properties (see p. 47).

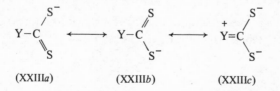

(XXIII*a*)   (XXIII*b*)   (XXIIIc)

The carboxylate anion is usually either monodentate for one metal or bidentate for two, but in a few examples such as hydrated zinc acetate (XXIV), it functions as a bidentate chelating ligand.

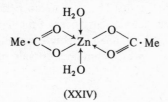

(XXIV)

Most other ligands which form 4-membered ring chelates are inorganic anions, e.g. SO_4^{2-}, CrO_4^{2-} and MnO_4^{2-}.

The general principles on which the structures of bidentate chelating ligands are based also apply to tridentate, quadridentate, and higher chelating systems. Thus, exactly the same types of donor groups are involved, and the ligand atoms are usually separated by sp^2- or sp^3-hybridised carbon atoms. The rings formed by chelation are nearly always 5- or 6-membered, a tridentate chelating ligand, for example, producing either a 5–5, 5–6, or 6–6 fused ring system, as illustrated by the copper(II) chelates (XXV), (XXVI) and (XXVII) respectively.

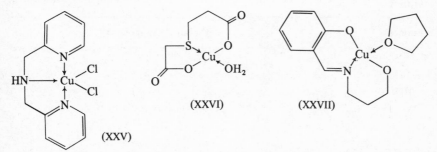

(XXVI)   (XXVII)

(XXV)

The presence of three or more donor groups in a ligand often allows the ligand to adopt more than one stereochemical arrangement around the chelated metal. Thus it is possible for diethylenetriamine and other flexible ligands which have a linear arrangement of their ligand atoms to co-ordinate so that the two resultant chelate rings are either coplanar or mutually at right angles, as illustrated in the general structures (XXVIII) and (XXIX) for an octahedral complex.

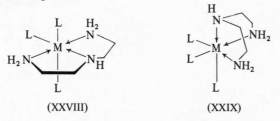

(XXVIII)   (XXIX)

In fact, a high proportion of the tri- and quadridentate ligands whose metal chelates are of interest are imines (Schiff bases) in which the nitrogen of the $-CH=N-$ group is one of the donor atoms. The rigid stereochemistry of the imine group usually allows only a coplanar arrangement of the ligand atoms as in (XXVII). Tridentate imines are obtained by condensation of an amine with a carbonyl compound, both of which contain a suitably placed donor atom.

$$\underset{/}{\overset{L'}{C}}=O + H_2N \underbrace{\qquad} L \longrightarrow \underset{/}{\overset{L'}{C}}=N \underbrace{\qquad} L + H_2O$$

Quadridentate di-imines are obtained similarly from suitable diamines or dicarbonyl compounds (see table 1.3).

The most important examples of planar quadridentate ligands are the porphyrins. These naturally-occurring macrocycles are formally derived from the parent porphin (XXX) by substitution of some or all of the β-hydrogens

Table 1.3. *Some amines and carbonyl compounds used to prepare chelating imines*

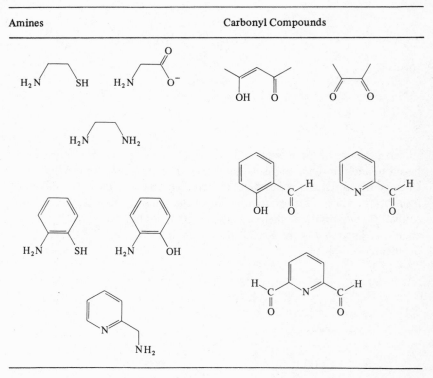

Amines	Carbonyl Compounds

on the four heterocyclic rings by substituents such as Me, Et, $-CH=CH_2$, and $-CH_2 . CH_2 . CO_2H$. Because both NH groups are appreciably acidic (both pK_1 and pK_2 are about 16), porphyrins function as weak dibasic acids, and with a variety of metal ions the dianion forms very stable chelates in which the four nitrogen atoms are arranged in a square-planar fashion about the metal. The iron(II) chelates (haems) play vitally important roles as the prosthetic groups of a number of haemoproteins such as the haemoglobins, cytochromes, and certain peroxidases. In most haemoproteins one or both of the remaining co-ordination positions of the iron are occupied by monodentate ligands (see XXXI) present in the protein. In haemoglobin, for example, one of the positions is occupied by the heterocyclic nitrogen atom of a globin histidine residue, and the other by a water molecule. The latter ligand is displaced by O_2 when the haemoglobin combines reversibly with oxygen to form oxyhaemoglobin (see pp. 34 and 69).

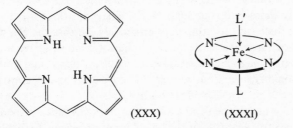

(XXX)   (XXXI)

Closely related to porphin and its derivatives are chlorin (XXXII) and corrin (XXXIII), the macrocyclic components of the chlorophylls and vitamin B_{12} respectively. In vitamin B_{12} the ligand is chelated with cobalt, in the chlorophylls with magnesium.

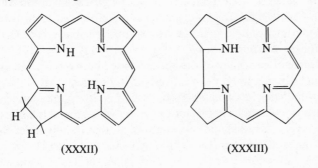

(XXXII)   (XXXIII)

1.3.   σ-Donor strength

All metal–ligand σ-bonds may be regarded as being formed by overlap of a filled orbital on the ligand with an empty orbital on the metal. Although such bonds are traditionally represented by a covalent symbol, as in the general structure $M-L$, they all have some ionic character as the result of the electronegativities of the metal and ligand atom being different. We can represent this situation by indicating that the bond is a resonance hybrid, as with a univalent metal cation and a univalent anion.

$$M-L \longleftrightarrow M^+ \; L^-$$

(XXXIVa)　　　　(XXXIVb)

In the extreme structure (XXXIVa) the metal is attached to the ligand by a purely covalent bond, whereas in the ionic structure (XXXIVb) only electrostatic forces are involved.

In general, the greater the difference between the electronegativities of the metal and the ligand atom the higher the degree of ionic character, i.e. the higher the contribution of the ionic structure (XXXIVb). Consequently the most ionic bonds are those between the highly electropositive metals (Rb, Cs) and the highly electronegative halogens (F, Cl). With less extreme combinations, particularly those involving small, highly charged cations which can strongly polarise the ligand and thus very effectively increase the electron density along the internuclear axis, the metal–ligand bond will have substantial covalent character. Even in a compound such as LiF, which would normally be regarded as 'ionic', calculations suggest that covalent bonding contributes about 25 per cent of the total bond strength.

With a particular metal, the degree of covalency of the metal–ligand σ-bond increases as the electronegativity of the ligand atom decreases, for example as one passes along the series of ligands NR_3, PR_3, AsR_3 and SbR_3, and along the anions F^-, Cl^-, Br^-, I^-, and F^-, HO^-, H_2N^-, H_3C^-.

The ability of a ligand to bestow covalent character on the σ-bonds which it forms with metals is often termed *σ-donor strength*, and a ligand which forms bonds having high covalent character is therefore said to be a strong σ-donor. Taking an example from the two series given above, both Cl^- and HO^- are stronger σ-donors than F^-.

As the degree of covalency of the bonds formed between a metal and a specific ligand depends upon the electronegativity of the metal, it may appear at first sight that when drawing up a sequence of relative donor strengths of a series of ligands it would be necessary to specify which metal is being considered. However it is usually not necessary to do this, as in practice the sequence is the same for all metals with only minor exceptions.

The concept of σ-donor strength is a useful one, for it is a measure of a ligand's ability to bestow covalent character upon the σ-bonds it forms with metals, and is therefore also a measure of its ability to decrease the positive charge on a metal cation. In other words, *σ-donor strength is a measure of a ligand's ability to increase the electron density on a metal through σ-bond formation*. As this electron density can determine the reactivity of a metal and of its ligands, it is profitable to consider some of the factors that control the σ-donor strengths of the various types of ligands.

As the electronegativity of an element is not an invariable property but is dependent upon the molecular environment, it follows that a series of ligands having the same donor atom, e.g. HO^-, PhO^- and H_2O, do not necessarily have the same donor strengths. The presence of a negative charge on the ligand atom, for example, reduces its electronegativity and hence increases σ-donor strength. The anionic ligands HO^- and RO^- are therefore both stronger σ-donors than H_2O and ROH, and the same is true for the analogous nitrogen and sulphur systems. Any reduction in electron density on the ligand atom as the result of mesomeric and inductive effects of substituents decreases donor strength. Resonance stabilisation therefore causes ArO^- and $ArNH_2$ to be weaker σ-donors than RO^- and $R.NH_2$ respectively, and the benzylic carbanions (XXXV; X = CH) to be weaker donors than H_3C^-. With the carbanions (XXXV), the expected order $(X = CH) > (X = N) > (X = {}^+NH)$ is observed. Resonance stabilisation similarly results in an amido nitrogen being a weaker donor than an amino one. As amides are resonance stabilised to a greater extent than esters, the carbonyl oxygen of an ester is a weaker donor than that of an amide.

(XXXV)

Although NH_3 and PH_3 are comparatively strong σ-donors, the fluorinated derivatives NF_3 and PF_3 are almost devoid of donor character. Similarly with other tri-substituted phosphines, donor strength decreases in the order $P(cyclohexyl)_3 > P(C_6H_5)_3 > P(C_6F_5)_3$, and $PEt_3 > P(OEt)_3 > PF_3$.

The state of hybridisation of the ligand atom is also important in that as s-electrons are, on average, closer to the nucleus and hence more strongly bound than p-electrons, they are less available for covalent bond formation. From this it follows that the greater the s-character of the orbital used for bonding to the metal the lower the σ-donor strength. This is in agreement with the observed orders shown below.

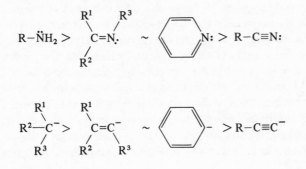

In fact a number of important ligands σ-bond to a metal by means of an sp-hybrid orbital on carbon, and the observed order of σ-donor strength is that predicted by consideration of the electronic effect of the atom attached to the carbon, i.e.

$$R-C\equiv C^- > N\equiv C^- > R-^+N\equiv C^- > {}^+O\equiv C^-$$

Having commented on the relative σ-donor strengths of different ligands, it is appropriate to indicate how these strengths are assessed experimentally.

We have seen that σ-donor strength is a measure of a ligand's ability to bestow covalent character upon the bonds which it forms with metals, or in other words, the ability of a ligand to transfer electron density into a metal orbital. This transference of electron density, or electron delocalisation, which results from the formation of a partially covalent bond manifests itself in a variety of ways, and by measuring suitable physical properties such as spectroscopic characteristics one can determine the degree of electron delocalisation associated with different ligands, and hence deduce relative σ-donor strengths. Quantitative studies of dipole moments, e.s.r. hyperfine splittings, n.m.r. chemical shifts, ionisation potentials, infrared frequencies, and spin orbital coupling parameters have all been used in this connection, and the reader is referred to some of the standard texts on co-ordination chemistry in which this topic is discussed more fully, e.g. Huheey (1972) and Basolo and Pearson (1967).

One very important point that must be emphasised at this stage is that *σ-donor strength should not be directly compared with σ-bond strength.* Indeed, there is often an apparent anomalous relationship between them. Thus although the σ-donor strengths of the halogens decrease in the order $I^- > Br^- > Cl^- > F^-$, the strength of σ-bonds between halogens and metals of Groups IA and IIA, for example, decrease in the reverse order, i.e. $M-F > M-Cl > M-Br > M-I$. Similarly, although oxygen compounds are usually weaker σ-donors than the corresponding sulphur compounds, with highly electropositive metals such as Mg^{2+} and Al^{3+} the M−O bond is stronger than the M−S one. These situations are directly analogous to that encountered when one compares the covalent character and the strength of the C−H bonds in ethane, ethylene and acetylene. Covalent character decreases as one passes along the series CH_3-H, $=CH-H$, $\equiv C-H$ (and here again this order is ascribed to the increasing s-character of the carbon orbital involved in the C−H bond), but the strength of the bond actually increases. The reason for this and similar anomalies is that the reduction in covalent character of the bond concerned is associated with a corresponding increase in ionic character. In the systems mentioned above the total combination of covalent and ionic bonding actually results in a bond which is stronger rather than weaker.

Generally speaking, σ-bond strength correlates directly with σ-donor strength only when a series of closely related ligands is being considered and where the σ-donor strength varies as the result of changes in the electron density on the ligand atoms, which must be the same - and in the same hydridisation state, of course - for all the ligands.

Related to this last point concerning covalent *versus* ionic bonding is the observation that with ligands which can form both σ- and π-bonds with a metal, the total strength of the metal–ligand bond is determined by both the σ- and π-components. With π-acceptor ligands any electronic effect which reduces the σ-donor strength generally causes a greater increase in the π-acceptor strength, and hence one often finds that some of the strongest metal–ligand bonds involve very weak σ-donors, e.g. PF_3 and $(CN)_2C=C(CN)_2$.

Before considering π-bonding between metals and ligands it is worthwhile briefly looking at some examples of the chemical consequences of the fact that different ligands have different σ-donor strengths.

A metal accepts electron density from all its ligands *via* the metal–ligand σ-bonds. Consequently the replacement of one or more of the ligands by a stronger σ-donor reduces the capacity of the metal to accept electron density from the remaining ligands, i.e. the metal becomes more electropositive. The bonds to the remaining ligands consequently become more ionic in character, with a corresponding increase in the electron density on their ligand atoms. This in turn is reflected in the basic and nucleophilic properties of these atoms.

Accordingly one finds that with solutions of organometallic species where the metal is co-ordinated with the solvent, the carbanion character of the organic residue is determined by the donor strength of the solvent.

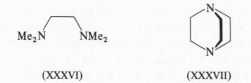

(XXXVI)   (XXXVII)

The basic and nucleophilic properties of organolithium compounds for example, are dramatically increased when 1,2-(dimethylamino)ethane (XXXVI, usually referred to as TMEDA, from *tetra*methyl*eth*ylene*dia*mine) is used as the solvent or co-solvent in place of the conventional diethyl ether (see American Chemical Society, 1974), and this allows many compounds that are unreactive in diethyl ether to be readily metallated.

$$Me_3Si \cdot CH_2Li \xleftarrow{Me_4Si} n\text{-BuLi/TMEDA} \longrightarrow \text{⟨benzene ring⟩–Li}$$

Although several factors are undoubtedly involved, this increase in reactivity can be largely ascribed to the replacement of the ether oxygens in the co-ordination sphere of the lithium by the stronger nitrogen donor ($R_3N >$ R_2O). The same effect can also be achieved by carrying out the metallations in the presence of other strong σ-donors such as t-butoxide anion, 1,4-diazabicyclo[2,2,2]octane (XXXVII), or triethylamine.

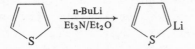

The reactivity of Grignard reagents in ether is similarly enhanced by the presence of pyridine.

For the same reason additional (axial) ligands on the metal are important in determining the reactivity of metalloporphyrins. In these macrocycles the size of the negative charge that resides on the chelated porphyrin dianion (see p. 11) is governed in the first instance by the electronegativity of the central metal, and with divalent metals is at a maximum with Ca^{2+} and Mg^{2+} Accordingly, these metals are often used to promote electrophilic attack on a porphyrin, e.g. in deuteriations and photochemical oxidations. However, even with these metals the reactivity of the porphyrin can be increased still further by putting strong σ-donors in the axial positions.

1.4. Carbon monoxide as a σ-donor/π-acceptor ligand

Largely as the result of the upsurge of interest in organometallic chemistry which commenced in the 1950s the number of known complexes in which carbon monoxide is present as one or more of the ligands has increased steadily since the first carbonyl, $Pt(CO)_2Cl_2$, was reported in 1870, and numerous carbonyls are now known for almost all of the metals in the transition block. These carbonyls have played a very important role in the development of co-ordination chemistry, particularly in our understanding of the nature of metal-ligand bonds, and it is not surprising that the various aspects of the chemistry of metal carbonyls have been extensively discussed in review articles, e.g. Abel and Stone (1969, 1970) and in books on organometallic and inorganic chemistry, e.g. Cotton and Wilkinson (1972). The aspects which are discussed in this book are mainly those which are of direct interest to the organic chemist, e.g. the susceptibility of co-ordinated carbon monoxide to nucleophilic attack, and the ability of carbon monoxide to stabilise low positive and negative oxidation states. The latter aspect is discussed in this section, and the former – together with synthetic applications – in sections 5.7 and 5.8.

Although the resonance structure $^-C{\equiv}O^+$ is often adequate for representing carbon monoxide in structural formulae, the co-ordinating properties of this compound are best explained by reference to the simple atomic orbital picture (XXXVIII) or its molecular orbital equivalent. The carbon and oxygen are

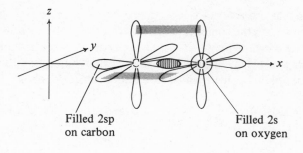

Filled 2sp
on carbon

Filled 2s
on oxygen

(XXXVIII)

linked by means of a σ-bond formed by overlap of a filled p-orbital on oxygen with an empty sp-orbital on carbon, and also by two π-bonds formed by lateral overlap of a pair of singly occupied p-orbitals on both carbon and oxygen. The picture is completed by an sp-orbital on carbon and a 2s-orbital on oxygen, both of which contain a lone pair of electrons. In the more accurate molecular orbital picture there are three filled σ-orbitals (corresponding to the σ-bond and the two lone pairs) and two filled degenerate π-orbitals, followed by the empty π-antibonding orbitals and a σ-antibonding orbital. Molecular orbital calculations reveal that the filled σ-orbital that corresponds to the oxygen lone pair is populated mainly by oxygen 2s-electrons. It is for this reason that in the atomic orbital picture (XXXVIII), the 2s- and 2p-orbitals on oxygen are not hybridised, as are those on carbon.

   In agreement with its high s-character, the lone pair on oxygen is held quite firmly, and it is the more weakly held lone pair in the sp-orbital on carbon that is responsible for the σ-donor properties of carbon monoxide. The compound is a weak base and under highly acidic conditions is protonated on carbon to give the formyl cation $H-C\equiv O^+$. This species is the active electrophile in the Gattermann–Koch synthesis of aldehydes from aromatic compounds, carbon monoxide, hydrogen chloride and a strong Lewis acid such as aluminium trichloride. The Lewis acid not only promotes the protonation of the carbon monoxide, but also enables what would otherwise be a thermodynamically unfavourable reaction to proceed to completion by forming a stable complex with the aldehyde.

$$HCl + CO + AlCl_3 \rightleftharpoons H-C\equiv O^+ + [AlCl_4]^-$$

$$H-C\equiv O^+ + ArH \underset{-H^+}{\rightleftharpoons} Ar-CHO$$

$$\big\updownarrow AlCl_3$$

$$\underset{\underset{H}{|}}{Ar-C=O}\longrightarrow AlCl_3$$

Carbon monoxide also acts as a weak σ-donor for Lewis acids, and with diborane reversibly gives the complex $BH_3 \cdot CO$. The complexes formed with trialkylboranes are of synthetic interest in that they are unstable and undergo a series of rearrangements involving alkyl migration from boron to carbon, and ultimately afford products which can be oxidised to tertiary alcohols.

$$R_3B^- - C \equiv O^+ \longrightarrow R_2B - \underset{\underset{O}{\|}}{C} - R \longrightarrow O=B \cdot CR_3 \xrightarrow{H_2O_2/HO^-} HO \cdot CR_3$$

The first of these boron to carbon migrations may be regarded as intra-molecular nucleophilic attack on co-ordinated carbon monoxide, and is identical to the rearrangement exhibited by certain metal carbonyls (p. 269).

Apart from the complexes formed with boranes all the known complexes of carbon monoxide – and they are numerous, both in variety and in type – are with transition metals. In the large majority of these carbonyls the carbon monoxide is bonded to one metal only as in $Ni(CO)_4$ and $Fe(CO)_5$, but in some it acts as a doubly or triply bridged ligand, i.e. it is bonded through its carbon atom to two and three metal centres respectively. Examples of important carbonyls that contain two distinct types of carbon monoxide ligands are nonacarbonyldi-iron (XXXIX), and octacarbonyldicobalt (XL), both of which also contain a metal – metal single bond. The first carbonyl is frequently used for preparing stable iron carbonyl complexes of organic ligands which in the unco-ordinated state would be exceedingly short lived (p. 92), while the second is involved in cobalt-catalysed hydroformylations (p. 283).

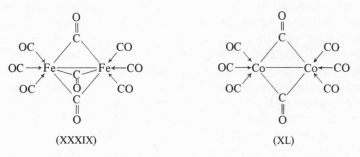

(XXXIX)   (XL)

The contrast between the very large number of known transition-metal carbonyls and the apparent non-existence of carbonyls of the main group metals is a striking one, and it is generally agreed that the disparity arises largely as the result of carbon monoxide being able to participate in back-bonding, i.e. act as a π-acceptor ligand. With transition metals, not only can the ligand be linked to the metal by a σ-bond (as would also be the case with main group metals), but a π-bond can also be formed by overlap of a filled d-orbital on the metal with an orbital on the carbon monoxide. With main group metals, of course, the d-electrons are too tightly held to participate in metal–ligand π-bonding. In the transition-metal carbonyls where the carbon monoxide is bonded to only one metal (and only this type is discussed here), the ligand orbital involved in the π-bond is one of the empty π-antibonding orbitals in the m.o. picture, or one of the carbon 2p-orbitals in the valence-bond picture (XLI).

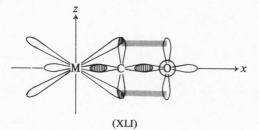

(XLI)

Even if the σ- and π-bonds existed independently of each other, one would expect the total metal–carbon bond strength to be higher with transition metals than with main group metals merely because of the existence of the π-bond. However, there is a synergistic interaction between the two types of bond and the σ-bond, for example, is strengthened by the presence of the π-bond. This is because the metal and ligand orbitals responsible for the σ-bond can overlap to a greater extent, for the increase in electron density on the metal that is associated with σ-bond formation is minimised by the backbonding. Similarly the π-bond is strengthened as a result of the electron density on the metal being increased by the σ-donor properties of the ligand. In other words, the total bond strength is greater than it would be if the σ- and π-components existed independently of each other.

Of all the σ-donor ligands that are known to exhibit π-acceptor properties, carbon monoxide is undoubtedly the most important, the most widely studied and, with the exception of PF_3 and CS, the strongest π-acceptor. It is appropriate, therefore, that the properties of this ligand be considered in some detail in order to make some general conclusions which apply to all ligands in the σ-donor/π-acceptor class.

According to the valence-bond interpretation, π-bonding in metal carbonyls involves a p-orbital on carbon as shown in (XLI). Clearly, the greater the extent to which this orbital is involved in π-bonding with the metal, the lower the extent it is involved in π-bonding with the oxygen. As back-bonding increases, therefore, the carbon–oxygen bond order decreases from three towards two, and in the extreme situation the bonding can be represented as shown in (XLII*a* and *b*).

On this basis it is clear that one is justified in representing carbon monoxide bonded to a formally zerovalent metal by the resonance structures (XLIII*a*–*c*). The first of these is consistent with the σ-donor properties of the ligand, the

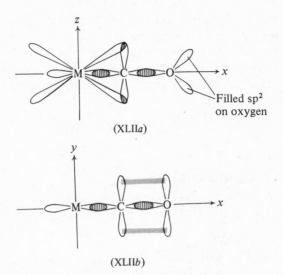

(XLII*a*)

(XLII*b*)

second takes its π-acceptor properties into account, while the third indicates the polarisation of the carbon–oxygen π-bond which results from the higher electronegativity of the oxygen.

$$\overset{-}{M}-C\equiv\overset{+}{O} \longleftrightarrow M=C=O \longleftrightarrow M=\overset{+}{C}-O^{-}$$

(XLIII*a*)   (XLIII*b*)   (XLIII*c*)

One very important point to notice concerning these resonance structures is that if one removes the CO entities with the carbon atoms in the closed shell (neon) configuration, i.e. as $\mathrm{C}\equiv\mathrm{O}^{+}$ and $^{2}\mathrm{C}=\mathrm{O}$ (or the polarised equivalent, $\mathrm{C}-\mathrm{O}^{-}$), the oxidation state of the metal is two units higher in the π-bonded structures (XLIII*b* and *c*) than in the solely σ-bonded structure (XLIII*a*). This is, in fact, a specific illustration of a general principle, i.e. *π-bonding between a metal and a π-acceptor ligand can always be represented in valence-bond terms by indicating that there is a contribution to the ground state by resonance structures in which the metal has undergone a two-electron oxidation*. Although these structures are associated with all the usual limitations of the valence-bond treatment of co-ordination compounds, they have the advantage that they can often be used for interpreting – and in some cases predicting – the properties of π-bonded ligands. They are particularly useful for indicating how back-bonding from the metal alters the electron distribution at different positions in the ligand. In the case of co-ordinated carbon monoxide, for example, structures (XLIII*b* and *c*) indicate that the CO residue acquires carbonyl-like character, and one can therefore predict that the donor properties of the oxygen are increased by the back-bonding. This prediction has recently been confirmed, in fact, by the discovery of complexes, e.g. (XLIV), in which the carbonyl oxygen acts as a Lewis base.

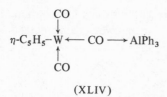

(XLIV)

As back-bonding alters the electron distribution – and hence the reactivity – of a co-ordinated ligand, it is clearly important to identify the factors that control the extent of back-bonding in metal complexes. Fortunately it is relatively easy to make a qualitative assessment of the relative degree of back-bonding to carbon monoxide for, as indicated above, a major consequence of this back-bonding is a reduction in the C—O bond order. This is manifested by a reduction in the C—O stretching frequency, which in free carbon monoxide is about 2155 cm^{-1}. Although the magnitude of this reduction is a good indicator of the extent of back-bonding, the data obtained, e.g. from infrared and Raman spectra, must always be interpreted with care for there are other factors besides back-bonding (see below) which reduce the C—O stretching frequency. However, from the spectroscopic data it has been deduced that the most important factors that control the extent of back-bonding are the electronic configuration and oxidation state of the metal, and the nature and stereochemical arrangement about the metal of the other ligands. In general, back-bonding increases, i.e. ν_{CO} decreases, as negative charge is added to the metal. Thus for a specific metal the back-bonding is most pronounced in the lowest oxidation state, and for a fixed electronic configuration, e.g. d^{10}, it is favoured by an increase in formal negative charge:

$Ni(CO)_4$	$[Co(CO)_4]^-$	$[Fe(CO)_4]^{2-}$	$[Mn(CO)_4]^{3-}$
2046	1890	1730	1670 cm^{-1}

Similarly, back-bonding is increased if the replacement of a ligand in a specific metal complex results in an increase in the electron density on the metal, e.g. if carbon monoxide is replaced by other ligands, for these are nearly always stronger σ-donors but weaker π-acceptors:

$Ni(CO)_4$	$Ni(PMe_3)(CO)_3$	$Ni(PMe_3)_2(CO)_2$
2046	1943	1934 cm^{-1}

Similarly, when a halogen is present as a ligand the degree of back-bonding correlates with its σ-donor strength, as in PtX(CO)(dipyr), where for X = Cl, Br and I, ν_{CO} = 2145, 2132 and 2120 cm^{-1}, respectively.

Back-bonding is decreased, of course, by the presence of a ligand which can reduce the electron density on the metal by acting as a π-acceptor. This is particularly true when the ligand is *trans* with respect to the carbon monoxide, for the two are then competing directly for the same pair of metal d_π-electrons.

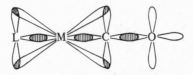

A good example of this situation is provided by zerovalent molybdenum complexes of type (XLV), where all the carbon monoxide ligands are *trans* to a phosphine. Now phosphines (PR_3) can act as π-acceptors by using an empty 3d-orbital on the phosphorus (see p. 32) and their π-acceptor strengths are dependent upon the electronegativity of the group (R) and, for example, increase in the order Ph $<$ Cl $<$ F. Back-bonding to the carbon monoxide ligands in (XLV) therefore *decreases* in the same order.

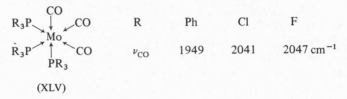

R	Ph	Cl	F
ν_{CO}	1949	2041	$2047\,cm^{-1}$

(XLV)

It must be pointed out at this stage that the decrease in C—O stretching frequency associated with an increase in the electron density on the metal is the overall result of at least two distinct mechanisms. The increase in negative charge on the metal not only causes an expansion of the d-orbitals and hence promotes π-bonding between the metal and the carbon monoxide, but it also reduces the extent to which the metal accepts electron density from the filled orbital responsible for the σ-donor properties of carbon monoxide. As this orbital is slightly antibonding in the free ligand, the net result is a weakening of the C—O bond and hence a reduction in the C—O stretching frequency. The implication, therefore, is that one must interpret differences in ν_{CO} values with care. Thus, the fact that carbonyls with $P(OR)_3$ ligands generally exhibit higher ν_{CO} values than the corresponding ones with PR_3 ligands does not necessarily mean that the former type of ligand is the stronger π-acceptor, for the observation can also be explained by the higher σ-donor strength of phosphines. This ambiguity does not invalidate, however, the general conclusion that back-bonding to carbon monoxide – and, indeed, to any π-acceptor ligand – is promoted by any factor that causes an increase in the electron density on the metal.

1.5. Other σ-donor/π-acceptor ligands

In table 1.4 are listed some of the other types of σ-donor ligands which can act as π-acceptors because they possess an empty or partly empty orbital which can overlap with a filled d-orbital on the metal. In the valence-bond picture this ligand orbital is a p- or d-orbital on the ligand atom for all the

Table 1.4. *Some σ-donor/π-acceptor ligands*

General type		Types of ligand
X≡Y	⁻C≡C−	Alkynyl anions
	⁻C≡N	Cyanide anion
	⁻C≡N⁺−	Isocyanides
	⁻C≡O⁺	Carbon monoxide
	⁻C≡S⁺	Thiocarbonyl
	:N≡C−	Cyanides
	:N≡N	Dinitrogen
	:N≡O⁺	Nitrosonium cation
X=Y	:O=C⟨	Aldehydes, ketones, carboxylic acid derivatives
	:O=O	Dioxygen
	:S=C⟨	Thiocarbonyl compounds
X⟨	:S⟨	Sulphides
	:P−⟨	Trialkyl-, triaryl-, and trihalo-phosphines; trialkyl- and triaryl phosphites
(ring X)	(ring ⁻)	Aryl anions
	(ring N)	Pyridine-type heterocycles
X=...=Y	−N=...=N−	Conjugated di-imines
	O...O	o-Quinones

types of ligands listed, but in the more sophisticated molecular orbital picture those unsaturated ligands that bond to the metal through a multiply bonded ligand atom use a π-orbital associated with the unsaturation. In all cases the filled metal orbital must have, of course, the correct size and symmetry if overlap is to occur; in an octahedral complex, for example, this means one of the t_{2g} (d_{xy}, d_{yz}, d_{xz}) class.

Within a specific class of ligand the π-accepting properties can vary tremendously from one ligand to another, but in general terms π-acceptor strength is increased by any modification which increases the electronegativities of those atoms on which negative charge accumulates as the result of back-bonding. These atoms can be identified most conveniently by drawing the appropriate resonance structures with the ligand bonded to a neutral species (M) to represent a zerovalent metal. This technique reveals, for example, that with an aryl anion it is the *ortho*- and *para*-carbons that are involved.

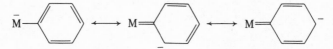

Accordingly, back-bonding is increased with this particular ligand if the hydrogens on these carbons are replaced either by more electronegative atoms such as halogens or by unsaturated groups ($-CN$, $-CO.R$) which can delocalise the negative charge still further.

For obvious electronic reasons, any modification that increases the π-acceptor strength of a specific ligand also causes to some extent a reduction in the σ-donor strength. Now the presence on a metal of any ligand that is a strong π-acceptor but a weak σ-donor effectively decreases the electron density on the metal and therefore reduces the ease with which the metal is oxidised. It is not surprising, therefore, that the various types of ligands listed in table 1.4 occur frequently in complexes where the metal is formally in a low positive or negative oxidation state. Some of the general chemical properties of these complexes are described in the following section (p. 38).

It should be noted that back-bonding takes place by a conjugative (mesomeric) electron displacement and involves electrons that are polarised fairly easily. Consequently, although in a specific metal complex there may only be a low degree of back-bonding to a particular ligand, this back-bonding will be increased under the influence of polarising forces. For example, it will be increased if a positive charge is generated in the ligand as the result of attack by an electrophile. This increased back-bonding would serve, of course, to stabilise the positive charge and hence lower the energy of the transition state of the reaction. The metal would therefore be functioning in the same way as those organic substituents, e.g. OH, NR_2, which activate aromatic systems towards electrophilic attack.

Having made a number of generalisations concerning π-acceptor ligands, we shall now illustrate these generalisations by reference to some specific types of ligand. Particular attention will be given to the way in which the physical and chemical properties of the different types are modified as the result of back-bonding from the metal.

A high proportion of the ligands listed in table 1.4 are isoelectronic with carbon monoxide and may be represented by the general formula $X \equiv Y$ where X, the ligand atom, is either C^- or N, and Y is either CR, N, ^+NR or O^+. As illustrated earlier in the discussion on carbon monoxide (p. 20), back-bonding with this type of ligand increases the electron density on the group Y, and hence it is not surprising that with a specific ligand atom the π-acceptor strength of the ligand reflects the electronegativity of the group Y and, in the case of $X = C^-$ for example, increases in the order $\bar{C} \equiv C- < \bar{C} \equiv N < \bar{C} \equiv \overset{+}{N}R < \bar{C} \equiv \overset{+}{O}$.

The increase in electron density on the group (Y) is obviously dependent upon the degree of back-bonding from the metal, and is manifested by a higher nucleophilicity and basicity. Accordingly, in contrast to organic 1-alkynes, certain σ-bonded metal derivatives of acetylene are sufficiently nucleophilic at the β-carbon atom to react with isocyanates to give the expected amide.

$$\underset{\displaystyle \eta\text{-}C_5H_5-Ni-C\equiv CH}{\overset{\displaystyle PPh_3}{\big\downarrow}} \xrightarrow{\text{O=C=N·Ph}} \underset{\displaystyle \eta\text{-}C_5H_5-Ni-C\equiv C-CO\cdot NH\cdot Ph}{\overset{\displaystyle PPh_3}{\big\downarrow}}$$

A wide variety of transition-metal cyanides can be readily alkylated on nitrogen to give the corresponding isonitrile. This reaction is involved in the traditional procedure for preparing organic isonitriles from silver cyanide and an alkyl halide.

$$Ag \cdot CN + RX \longrightarrow X-Ag \longleftarrow \bar{C} \equiv \overset{+}{N} \cdot R \xrightarrow{2KCN} \bar{C} \equiv \overset{+}{N} \cdot R + K^+[Ag(CN)_2]^- + KX$$

It should be noted that with free cyanide anion (and hence with alkali-metal cyanides) alkylation occurs predominantly on carbon to give the alkyl cyanide (cf. p. 132).

Hydrogen cyanide is a weak base and strongly acidic conditions have to be employed for protonation to occur to an appreciable extent, as in the Gattermann synthesis of aromatic aldehydes (cf. Gattermann-Koch reaction, p. 17).

$$H-C\equiv N \xrightarrow{HCl/ZnCl_2} \underset{\underset{\displaystyle +}{\displaystyle H-C=NH}}{\overset{\displaystyle H-C\equiv \overset{+}{N}H}{\updownarrow}} \xrightarrow{\text{Ar H}} \underset{\displaystyle NH}{\overset{\displaystyle Ar \cdot C-H}{\|}} \xrightarrow{H_2O} \underset{\displaystyle O}{\overset{\displaystyle Ar \cdot C-H}{\|}}$$

In contrast, certain transition-metal cyanides are protonated comparatively readily. Complexes of type (XLVI), for example, can be successfully titrated with perchloric acid and converted into stable salts by dissolution in mineral acids.

$$M(phen)_2 (C\equiv N)_2 \xrightarrow{\quad 2H^+X^- \quad} [M(phen)_2 (C\equiv\overset{+}{N}H)_2]\ 2X^-$$

(XLVI)

M = Fe, Ru, Os.

With this particular type of complex the basicity of the cyanide nitrogen increases in the order Fe < Ru < Os, which reflects the relative degrees of back-bonding with these metals.

Back-bonding is similarly responsible for the enhanced basicity of the co-ordinated dinitrogen in those complexes which contain an M←N≡N unit, e.g. ReCl(N$_2$)(PMe$_2$Ph)$_4$ where the basicity of the β-nitrogen is of the same order as that of the oxygen in diethyl ether. Protonation of co-ordinated dinitrogen is currently thought to be an important step in the biological conversion of molecular nitrogen into ammonia by the nitrogen-fixing enzyme nitrogenase (see the book on nitrogen fixation, edited by Postgate, 1971). Part of this enzyme is a molybdenum–iron protein and one scheme which may be visualised for the formation of ammonia involves co-ordination of dinitrogen with molybdenum followed by stepwise protonation of the ligand and associated oxidation of the metal.

$$Mo \longleftarrow N\equiv N \xrightarrow{\ H^+\ } \overset{+}{Mo}-N=NH \xrightarrow{\ H^+\ } \overset{2+}{Mo} \longleftarrow \underset{\underset{H}{|}}{N}=NH$$

$$\Big\downarrow H^+$$

$$\overset{4+}{Mo}=NH \xleftarrow{\ -NH_3\ } \overset{3+}{Mo}-\overset{+}{N}-NH_3 \xleftarrow{\ H^+\ } \overset{3+}{Mo} \longrightarrow \underset{\underset{H}{|}}{N}-NH_2$$

$$\Big\downarrow H_2O$$

$$\overset{4+}{Mo}=O + NH_3$$

Indeed, treatment of Mo(N$_2$)$_2$ (PMe$_2$Ph)$_4$ with sulphuric acid in methanol actually results in up to 20 per cent of the nitrogen content being released as ammonia.

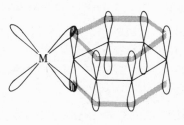

(XLVII)

One of the factors which are thought to be responsible for the greater thermal stability of σ-bonded transition-metal aryls compared with analogous metal alkyls is that in the former class of compound the metal–carbon bond is strengthened as the result of a π-bond component. This is formed by overlap of a filled d-orbital on the metal with a delocalised π-orbital on the ring (see valence-bond picture (XLVII)), and the appropriate resonance structures (see p. 24) predict that the negative charge transferred to the aryl group is shared by the *ortho-* and *para*-carbons. Confirmation of this prediction is provided by chemical and spectroscopic evidence. Thus, transition-metal aryls such as (XLVIII) undergo electrophilic substitution at the *para*-position (the *ortho*-position is sterically hindered), while the ^{19}F n.m.r. shielding parameters of a series of platinum(II) complexes (XLIX) and the corresponding *meta*-isomers show that the electron density is higher at the *para*-position than at the *meta*, and that the π-donor properties of the metal are highest when the ligand (L) is a strong σ-donor with low π-accepting capacity, e.g. CH_3.

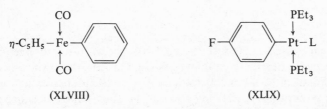

(XLVIII)   (XLIX)

However, it is evident from the data obtained from a variety of sources that unsubstituted aryl groups are comparatively poor π-acceptors, and when they have to compete with a number of strong π-acceptors, as in $ArMn(CO)_5$ for example, the degree of back-bonding is negligible. This is also true with those aromatic nitrogen heterocycles of the pyridine type which function as σ-donors by using the lone pair of electrons on nitrogen and as π-acceptors by using a delocalised orbital on the ring.

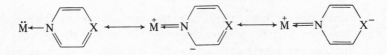

The π-accepting properties of these heterocycles increases (but σ-donor strength decreases) as the electronegativity of the group (X) is increased, e.g. as one passes along the series pyridine (X = CH), pyrazine (X = N), and pyrazinium cation (X = $^+$NH), and also when one introduces electron-withdrawing substituents into the 4-position of the pyridine system. This last point is illustrated by the values of ν_{CO} (see p. 21) for complexes of the type (L), in which the heterocyclic ligand has to compete with the carbon monoxide for the d_π-electrons on the metal.

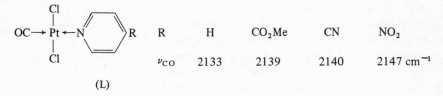

(L)

The π-accepting properties of these pyridine-type heterocycles are also increased by the introduction of electron-withdrawing substituents into the 2- and 6-positions, but this modification also causes severe steric interaction with those ligands in the *cis*-position on the metal.

On account of the greater delocalisation of charge, isoquinoline is a stronger π-acceptor than pyridine.

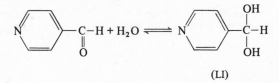

(LI)

Although there is ample spectroscopic evidence with transition metals of the first series (particularly divalent cobalt and nickel), the most convincing demonstrations of back-bonding to pyridine-type ligands has been provided by studies with ruthenium(II) and osmium(II). Thus in neutral, aqueous solution 4-formylpyridine exists as the hydrate (LI) to an extent of about 45 per cent, but co-ordination to a $[(NH_3)_5Ru]^{2+}$ entity, however, reduces this figure to less than 10 per cent because back-donation from the metal stabilises the carbonyl form.

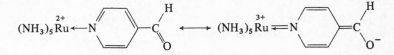

Evidently this stabilising effect more than compensates for the destabilising effect associated with the positive charge introduced into the heterocycle by co-ordination with a metal cation. However, this is not so when the oxidation state of the metal is increased, for co-ordination to $[(NH_3)_5Ru]^{3+}$ increases hydrate formation to over 90 per cent. The figure for co-ordination to $[Cr(NH_3)_5]^{3+}$ is similarly well over 90 per cent.

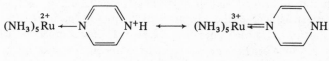

(LII)

Back-bonding to the pyrazinium cation (see (LII)) decreases the acidity of this species, i.e. the basicity of pyrazine is increased by co-ordination. In quantitative terms the pK_a of (LII) is 2.5 compared with 0.6 for the unco-ordinated pyrazinium cation. The effect of back-bonding is even more dramatic with the osmium(II) complex, $[^+OsCl(NH_3)_4(pyrazine-H)^+]$, which has a pK_a of 7.6. Understandably, replacement of the chloride anion in this complex by the comparatively strong π-acceptor, dinitrogen, reduces the pK_a to 0.3; i.e. in this case the basicity of pyrazine is decreased slightly by co-ordination.

An increase in the pK_a value by five units has been reported to occur when monoprotonated p-benzoquinonedi-imine is co-ordinated with a $[Ru(NH_3)_5]^{2+}$ entity (see (LIII)).

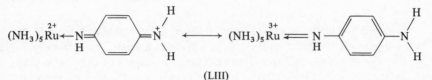

(LIII)

Certain conjugated bidentate ligands such as aliphatic di-imines (LIV), o-benzoquinonedi-imines (LV), and related compounds in which one or both NR groups have been replaced by O or S, function as very efficient π-acceptors.

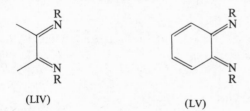

(LIV)   (LV)

In such ligands the π-orbitals associated with the unsaturation extend over both ligand atoms, and can therefore overlap with a filled orbital on the metal to form a cyclic, delocalised 6π-electron system. The metal complexes of the ligands are therefore electronically analogous to pyrrole and related five-membered heterocycles.

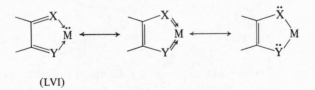

(LVI)

Because of the extensive interaction of metal and ligand orbitals in these complexes it is often impossible to assign an oxidation state to the metal, but one general observation that can be made is that the contribution of structures of type (LVI) to the ground state is reduced as the electron density on the metal increases. The contribution is minimal, for example, with zerovalent metals such as Pd(0), and Pt(0), and when *o*-benzoquinone is treated with a source of Pd(PPh$_3$)$_2$ the bond lengths of the resultant complex are far more consistent with the palladium(II) formulation (LVII*b*) than with the palladium(0) one (LVII*a*).

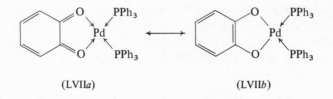

(LVII*a*)   (LVII*b*)

Associated with the extensive intermixing of metal and ligand orbitals is the characteristic ability of these ligands to form a series of structurally similar complexes which differ from each other in respect of the charge they carry (see review by McCleverty, 1968). These series are formed by stepwise, reversible, one-electron oxidations, as illustrated by the progressive oxidation of the dianion (LVIII) to the dication (LIX).

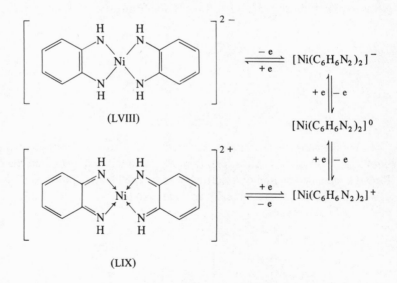

If the five species in this particular series are all formally regarded as nickel(II) complexes, the various oxidation steps correspond to the stepwise conversion of both ligands from the dianion stage (LX) to the di-imine stage (LXI). However, in view of the nature of the metal-ligand bonding in such complexes it is probably unwise to distinguish between oxidation of the ligands and oxidation of the metal.

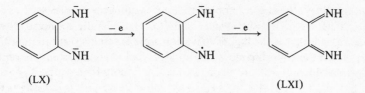

(LX) (LXI)

Closely related to these oxidations are the dehydrogenations which readily occur when ligands of the 1,2-diaminoethane type are co-ordinated with Ru^{2+} and Fe^{2+}.

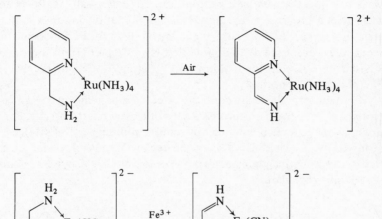

As hydrolysis of the imine-type ligands produced in these dehydrogenations affords the parent carbonyl compounds, these reactions appear to have considerable synthetic potential.

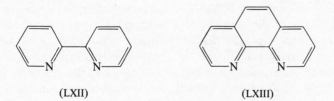

(LXII) (LXIII)

Structurally analogous to the conjugated bidentate ligands described above are the aromatic heterocycles dipyridyl (LXII) and 1,10-phenanthroline (LXIII), both of which function as fairly effective π-acceptors. One important piece of evidence which illustrates this is their ability to stabilise low oxidation states. This stabilisation is revealed in quantitative terms by the effect of these ligands on the reduction potentials (E^0) of metal couples such as Fe(III)/Fe(II), Ru(III)/Ru(II) and Cu(II)/Cu(I). Generally speaking E^0 values are decreased, that is the higher oxidation state is stabilised more than the lower, when the co-ordinated water molecules in a redox system are replaced by ligands which overall are stronger electron donors than water. Typical of such ligands are NH_3, HO^- and F^-, and their effect on E^0 is consistent with the increased ease with which a metal in a lower oxidation state loses an electron as the electron density on the metal is increased. In contrast, the replacement of co-ordinated water by dipyridyl, phenanthroline, or other weak σ-donors that also have π-accepting properties usually results in an increase in E^0 because of the greater stabilisation of the lower oxidation state (see also p. 38). Typically, the figure of +0.77 for Fe^{3+} (aqueous)/Fe^{2+} (aqueous) is decreased to +0.36 for $[Fe(CN)_6]^{3-}$/$[Fe(CN)_6]^{4-}$ but raised to +1.10 and +1.12 volts for $[Fe(dipy)_3]^{3+}$/$[Fe(dipy)_3]^{2+}$ and $[Fe(phen)_3]^{3+}$/ $[Fe(phen)_3]^{2+}$ respectively.

If suitable complexes are used, this effect of π-bonding on E^0 can be used to assess the relative π-acceptor strengths of closely related ligands. For example, the reduction potentials of couples of the type $[Ru(NH_3)_5L]^{3+}$/ $[Ru(NH_3)_5L]^{2+}$ where L is a six-membered nitrogen heterocycle, confirm that π-acceptor strength increases in the order pyridine < pyrazine < methyl 4-pyridinecarboxylate.

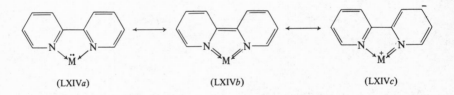

(LXIV*a*)   (LXIV*b*)   (LXIV*c*)

As with all π-acceptor ligands, back-bonding to dipyridyl can be represented by means of resonance structures. These include (LXIV*b*) and a total of ten dipolar structures such as (LXIV*c*). The last group indicate that the negative charge transferred to the ligand by back-bonding is shared by all ten carbons. In all the resonance structures that can be drawn, however, the aromaticity of one or both heterocyclic rings is lost, and evidently delocalisation of the electrons on the metal can only occur at the expense of the delocalisation in the rings. A similar situation exists with phenanthroline, and hence it is not surprising that the π-acceptor properties of both these ligands are much lower than that of the monocyclic analogue *o*-benzoquinonedi-imine (LV; R = H) where back-bonding actually increases the aromaticity of the six-membered ring (see top of p. 33).

Thermodynamic and spectroscopic data obtained from studies with transition-metal complexes of phosphines and phosphites have frequently

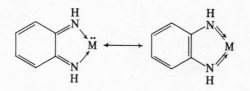

been interpreted in terms of the donation of electrons by the metal into the empty 3d-orbital on phosphorus.

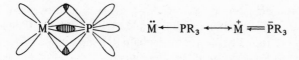

The data are consistent with this back-bonding increasing in the expected order $P(alkyl)_3 < P(aryl)_3 < P(OR)_3 < PCl_3 < PF_3$. The high π-acceptor properties of the last ligand, which is a very weak σ-donor and does not even form complexes with the boron trihalides, are illustrated by its ability to stabilise low oxidation states – and indeed even formally negative ones as in $[Co(PF_3)_4]^-$ and $[Fe(PF_3)_4]^{2-}$ (see review by Kruck, 1967). It must be pointed out, however, that the concept of back-bonding to phosphorus has caused a great deal of controversy in the past, and several eminent chemists have maintained that the data obtained from complexes of triaryl and trialkyl phosphines, for example, can be explained by alternative concepts such as changes in σ-bonding (see p. 22). This topic has been critically reviewed by Verkade (1972) and Mason and Meek (1978).

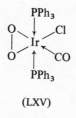

(LXV)

Dioxygen (O_2) is a π-acceptor ligand of obvious biological importance, and with transition metals it forms several types of complexes. Those having a metal : O_2 ratio of 1 : 1 have the metal bonded either to both oxygen atoms, as in the complex (LXV) formed by the oxidative addition of dioxygen to Vaska's compound (see p. 251), or to one oxygen atom. Complexes of the latter type include those formed reversibly from dioxygen and cobalt(II) chelates of certain square-planar quadridentate Schiff bases in which one of the axial co-ordination positions on the metal is occupied by a basic nitrogen atom.

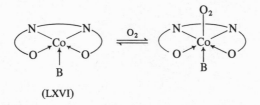

(LXVI)

These cobalt complexes serve as very useful models in current attempts to understand the chemical behaviour of naturally-occurring oxygen-carrying complexes such as haemoglobins (see reviews on these models by Basolo, Hoffman and Ibers, 1975, and McLendon and Martell, 1976). On the basis of e.s.r. results they are formally regarded as cobalt(III)–superoxide systems $(Co . O_2^-)$ with about 90 per cent of the spin density of the unpaired electron originally on the cobalt being transferred to the dioxygen molecule on co-ordination. In the Pauling model the $Co–O_2$ bonding is explained in terms of σ-donation from a non-bonding sp^2 lone pair on oxygen with back-bonding from filled d-orbitals on the metal $(3d_{xz}$ or $3d_{yz})$ into the half-empty π-anti-bonding orbitals on the O_2 entity. As the σ-donor strength of dioxygen is even lower than that of carbon monoxide, on the basis of the Pauling model the strength of the $Co–O_2$ bond is determined almost entirely by the extent of the back-bonding. In agreement with this one finds that the affinity for dioxygen of the metal in the pentaco-ordinate complexes (LXVI) is dependent on the electronic structures of the equatorial Schiff base ligand and of the axial base in the respect that it is increased by changes which increase the electron density on the metal.

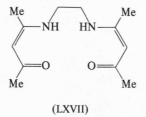

(LXVII)

With the Schiff base (LXVII), for example, the affinity for oxygen increases as one passes along the series of nitrogen bases: 4-cyanopyridine, pyridine, 4-methylpyridine, piperidine; i.e. as σ-donor strength is increased but π-acceptor strength is decreased. The affinity for oxygen is much higher when the base is imidazole than when it is pyridine, presumably due to the relatively strong π-donor properties of the former heterocycle.

1.6.   Ligands that form π-complexes with metals

All the ligands described in §§1.2–1.5 form metal–ligand σ-bonds by means
of filled atomic orbitals on specific ligand atoms. There are, however, several
types of unsaturated ligand (see table 1.5) which bond with a metal by using
the delocalised π-electrons associated with the unsaturation, and may therefore be considered to be bonded by two or more contiguous ligand atoms.
Because of their use of π-bonded electrons such ligands are said to form π-
(or η-) complexes (see p. 3). In those metal complexes of these ligands
where all the unsaturated carbon atoms are formally bonded to the metal
all the carbon atoms are in the same plane, with the metal lying above this

Table 1.5. *Ligands which use their π-electrons for bonding to a metal*

Number of π-electrons used for bonding	Ligand		
2	$\begin{array}{c}\backslash\quad/\\ C{=}C\\ /\quad\backslash\end{array}$ Alkenes, allenes	$\begin{array}{c}\backslash\quad+/\\ C{=}N\\ /\quad\backslash\end{array}$ Alkylated imines	
	$-C{\equiv}C-$ Alkynes	$-C{\equiv}N$ Nitriles	
4	Allyl anion	Cyclic and acyclic conjugated dienes	Cyclohexatrienyl cation
6	Cyclopentadienyl anion	Arenes	Cyclohepatrienyl cation
10	Cyclo-octatetraenyl dianion		

plane and equidistant from each carbon atom. Thus in ferrocene (LXVIII), all ten iron-carbon distances lie in the range 201–207 pm.

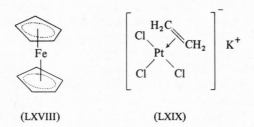

(LXVIII)   (LXIX)

It can be seen from table 1.5 that all these unsaturated ligands can be classified according to the number of π-electrons which they use for bonding to a metal. One point that must be emphasised in connection with this classification, however, is that the charges placed on the various ligands – and hence the number of π-electrons these ligands use for bonding – are assigned on a purely formal basis. Thus, it is largely a matter of personal choice whether one regards ferrocene (LXVIII), for example, as a combination of Fe^{2+} and two $(C_5H_5)^-$ anions (six π-electrons), or a combination of Fe^0 and two $(C_5H_5)\cdot$ radicals (five π-electrons). As indicated by the table, the author prefers the former viewpoint, but other authors (e.g. Coates, Green and Wade, 1968) prefer the latter.

In addition to acting as electron donors for a metal all the ligands in table 1.5 can accept electrons by using their π-antibonding orbitals. Indeed, in arene complexes such as bis(η-benzene)chromium(0) the overall electron drift is towards the ligands, which accordingly bear a substantial negative charge. Although there is hardly any information available concerning their relative π-acceptor strengths, it is clear that the degree of back-bonding to any of these ligands is influenced by the same factors – and in the same manner – that control the degree of back-bonding to all the π-acceptor ligands discussed in earlier sections. This is illustrated below for the most extensively studied class of π-bonded ligand, the alkenes.

Alkenes form complexes with a variety of transition metals (see reviews by Quinn and Tsai (1969), Nelson and Jonassen (1971), and Petitt and Barnes (1972)). Indeed, some of these complexes are amongst the oldest organometallic compounds known, the so-called Zeise's salt (LXIX) having been first described in 1830. Silver(I) complexes of alkenes have frequently been used in organic chemistry for purifying alkenes (cf. p. 82), and as suitable solid derivatives in structural determinations by means of X-ray crystallography.

According to the currently accepted theory, a metal–alkene bond contains a σ-component which arises from overlap of the filled π-bonding orbital of the carbon–carbon double bond with an empty orbital on the metal, and also a π-component arising from overlap of a filled metal orbital with the empty

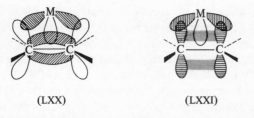

(LXX)   (LXXI)

π-antibonding orbital of the double bond (see (LXX)). The corresponding valence-bond picture (LXXI) is identical except that the filled metal orbital overlaps with the carbon 2p-orbitals responsible for the π-component of the double bond. A consideration of the latter picture suggests that any factor that increases the transfer of electron density from the filled metal orbital into the two carbon 2p-orbitals will decrease the overlap of these orbitals and this in turn will make less electron density available for the empty orbital on the metal. The same conclusion can be drawn from the molecular orbital picture. On this basis one is justified in representing any metal–alkene complex as a resonance hybrid of two extreme structures (LXXII*a*) and (LXXII*b*).

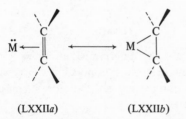

(LXXII*a*)   (LXXII*b*)

One result of the structure (LXXII*b*) contributing to the ground state is that on co-ordination the carbon–carbon 'double' bond in the alkene is lengthened (from 133 to 146 pm in (LXXIII)), and the hybridisation state of the two carbons of that bond moves from sp^2 towards sp^3. This change in hybridisation is reflected not only spectroscopically, e.g. by a decrease in the *cis* and *trans* proton–proton coupling constants in mono- and di-substituted alkenes, but also by the stereochemistry of the ligand. In the unco-ordinated state the two carbons of the double bond and the four substituent atoms are all in the same plane, but on co-ordination the substituent atoms move out of this plane and away from the metal (cf. structure LXXIV).

The degree of back-bonding from the metal, i.e. the extent to which structure (LXXII*b*) contributes to the ground state, is determined by the electron density on the metal and the electron-accepting properties of the alkene. Any factor that raises either the electron density on the metal or the electron-accepting properties of the alkene increases the degree of back-bonding. Accordingly, with isoelectronic metals bearing similar ligands, back-bonding increases with decreasing formal charge on the metal, e.g.

Hg(II) < Au(I) < Pt(0). Indeed, with zerovalent metals such as Pt(0), Pd(0) and Ni(0), the degree of back-bonding is sufficiently high to justify the routine use of the metallocycle type of structure (LXXII*b*) for representing their alkene complexes, even when the alkenes have only low electron-accepting properties, e.g. ethylene (see LXIII). With higher oxidation states such as Pt(II), Pd(II), Rh(I) and Ir(I), it requires strong π-electron-withdrawing substituents (CN, CO_2R) on the alkene before the degree of back-bonding justifies this representation (see LXXIV). In the case of mercury(II) complexes of alkyl-substituted alkenes, the degree of back-bonding is so low that the alternative representation (LXXII*a*) is far more appropriate. Certainly the overall transfer of electron density appears to be from the alkene to the mercury, for the co-ordinated alkene functions as a highly electrophilic centre (see p. 195).

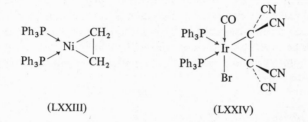

(LXXIII)   (LXXIV)

The electron density on the metal is also determined, of course, by the other ligands present. Ligands which are strong σ-donors but weak π-acceptors (amines, alkyl groups) raise the electron density and therefore increase the degree of back-bonding to the alkene.

The general observations that can be made concerning metal–alkene complexes, therefore, are that back-bonding is promoted by:

(*a*)   the metal being in a low oxidation state,

(*b*)   strong σ-donor but weak π-acceptor ligands on the metal, and

(*c*)   electron-withdrawing substituents on the alkene.

These factors are, of course, the same ones that promote back-bonding to any ligand that has π-accepting properties.

1.7.   Metals in low oxidation states

So far the two fundamental properties that have been discussed for the various types of ligands are σ-donor strength and π-acceptor strength. Now the former is a measure of a ligand's ability to transfer electron density to a metal *via* a σ-bond, while the latter is a similar measure of its ability to accept electron density from a metal *via* a π-bond. It follows from this that the attachment to a metal of ligands that are weak σ-donors but strong π-acceptors will reduce the electron density on the metal, and hence stabilise the lower oxidation states of that metal. Accordingly, the various types of π-acceptor ligands discussed in §§ 1.5 and 1.6 feature very strongly in the chemistry of those complexes in which the metal is *formally* in a low positive, zero, or negative oxidation state. Some selected properties of such complexes will now be briefly considered.

It is appropriate to point out first of all that the word 'formally' is emphasised in the above reference to oxidation states because in most cases the *actual* oxidation state of the metal is higher than is implied by the chemical structures used to represent such complexes and from which formal oxidation states are deduced. This discrepancy arises simply because in these chemical structures the ligands are invariably shown as σ-donors only, and no indication is given that they also accept electron density from the metal. The appropriate resonance structures can always be drawn of course, and these not only indicate a higher oxidation state for the metal but – as pointed out earlier – they are often useful for understanding and predicting the properties of the ligands. An advantage of showing only the σ-donor properties of the ligands, however, is that one can rapidly apply the Sidgwick effective atomic number (EAN) rule. In this rule it is considered that each of the ligand atoms in a metal complex donates two electrons to the metal which acts only as an electron acceptor, and that stability is associated with the metal having the same total number of electrons as the next rare gas. The rule has many exceptions (see Tolman, 1972), but is very often obeyed by organometallic compounds, especially the metal carbonyls. Thus in the stable binary carbonyls $M(CO)_n$ where the metal is formally zerovalent, the EAN of the metal (atomic number + $2n$) is almost always the atomic number of the next inert gas, e.g. 36 (krypton) in $Ni(CO)_4$, $Fe(CO)_5$ and $Cr(CO)_6$, and 86 (radon) in $Pt(CO)_4$, $Os(CO)_5$ and $W(CO)_6$.

The Sidgwick rule is also applicable to anionic metal carbonyls where the metal is formally in a negative oxidation state, e.g. $[V(CO)_6]^-$, $[Mn(CO)_5]^-$ and $[Fe(CO)_4]^{2-}$. The preparation and various reactions of these species have been discussed by King (1964, 1970) and Bruce and Stone (1968) amongst others. In most cases they can be prepared by the reductive cleavage of the metal–metal bonds in polynuclear metal carbonyls. Sodium amalgam in tetrahydrofuran is frequently used for the reduction:

$$Mn_2(CO)_{10} + 2Na \rightarrow 2\,Na^+\,[Mn(CO)_5]^-$$

$$[\eta\text{-}C_5H_5 \cdot Fe(CO)_2]_2 + 2Na \rightarrow 2\,Na^+\,[\eta\text{-}C_5H_5 \cdot Fe(CO)_2]^-$$

Some of these anions show remarkable stability, and can function as leaving groups in organic reactions, e.g. nucleophilic substitutions (see top of p. 40).

Nucleophilic displacement of an anionic metal carbonyl is one of the routes by which acyl residues are cleaved from the metal in certain carbonylations (see p. 275).

Associated with the high stability of some of these anionic carbonyls is

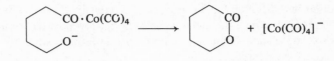

$$Me_3Si \cdot Co(CO)_4 + NMe_3 \longrightarrow [Me_3Si \cdot NMe_3]^+ \, [Co(CO)_4]^-$$

the high acidity of their conjugate acids. For $HCo(CO)_4$, a toxic unstable gas, the $pK_a \simeq 1.0$, and therefore in solution this hydride is a strong acid. For $H_2Fe(CO)_4$ $pK_1 \simeq 4.4$ and $pK_2 \simeq 14$. As expected, the replacement of the carbon monoxide entities by ligands that are stronger σ-donors but weaker π-acceptors decreases the acidity of these and related hydrides because they decrease the stability of the corresponding anion. For $HCo(CO)_3(PPh_3)$, for example, $pK_a \simeq 7.0$, while $HCo[P(OMe)_3]_4$ is such a weak acid that the hydrogen cannot be extracted as a proton by HO^- or MeO^- but requires the action of sodium hydride.

This reversible ionisation of certain transition-metal hydrides suggests that metals in low oxidation states are potentially basic in nature. This is indeed so, and the protonation of the metal in many complexes is now well documented (see Green and Jones, 1965, and Sc.·.iver, 1970). The ease of protonation varies greatly, and while some metals are very basic and protonate readily others require strong acids under anhydrous conditions.

$$Fe[P(OMe)_3]_5 \xrightarrow{(NH_4)^+ (PF_6)^-} H-\overset{+}{Fe}[P(OMe)_3]_5(PF_6)^- + NH_3$$

$$Ni[P(OEt)_3]_4 \xrightarrow{HCl \text{ in benzene}} H-\overset{+}{Ni}[P(OEt)_3]_4 \, Cl^-$$

$$Fe(CO)_5 \xrightarrow{Liquid \, HCl} H-\overset{+}{Fe}(CO)_5 \, Cl^-$$

Factors that determine the ease of protonation include the metal's ability to accommodate the increase in co-ordination number associated with the protonation, and the σ-donor and π-acceptor strengths of the ligands. With all the metals that can be protonated the formation of the metal–hydrogen bond can be detected by the characteristic high field ^1H n.m.r. signal in the range $\tau = 12$ to $\tau = 60$.

Many transition-metal complexes of metals in low oxidation states also exhibit Lewis basicity (see review by Kotz, 1969), and with Lewis acids such as the halides of mercury(II), zinc(II), gallium(III) and indium(III), give adducts that contain metal–metal bonds.

$$\eta\text{-}C_5H_5\!\!-\!\!\underset{\underset{PPh_3}{\uparrow}}{\overset{\overset{CO}{\downarrow}}{Co}}\!\longrightarrow\! MCl_2 \qquad\qquad [(CO)_4Co\longrightarrow GaBr_3]^-$$

$$M = Hg, Zn$$

Much more important, however, is the ability of these complexes – particularly the anionic ones – to act as nucleophiles and to displace from saturated carbon leaving groups such as halide, sulphonate and perfluorophenoxide anions. Because of the bulky nature of the nucleophile involved these displacements tend to be restricted to primary carbon atoms.

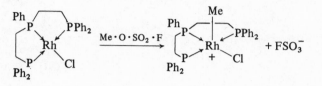

$$[\eta\text{-}C_5H_5\cdot Fe(CO)_2]^- \xrightarrow{\text{Et}\cdot\text{I}} \eta\text{-}C_5H_5\cdot Fe(CO)_2Et$$

Nucleophilic substitutions of this type have proved very useful for preparing alkyl cobaloximes (LXXV). These cobalt complexes of dimethylglyoxime have been extensively investigated (see review by Schrauzer, 1976) for they are structurally analogous to vitamin B_{12} in the sense that in both systems the cobalt is bonded to four nitrogen atoms arranged in a square-planar manner, and is also σ-bonded to a carbon atom and to a nitrogen base (B) such as pyridine or imidazole.

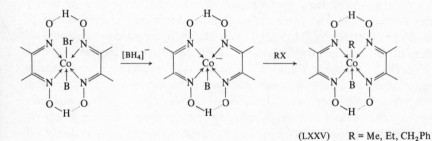

(LXXV)   R = Me, Et, CH$_2$Ph

The limited amount of stereochemical information available indicates that these nucleophilic substitutions proceed by an S_N2 type of mechanism with inversion of configuration at carbon. When iron is the nucleophilic centre, cleavage of the resultant metal–carbon bond by bromine is also accompanied (> 90 per cent) by inversion (see equation on p. 42).

The nucleophilicity of transition-metal complexes varies greatly, as shown by the following figures deduced from the relative rates of reaction of a range of anionic carbonyls with alkyl halides.

$[\eta\text{-}C_5H_5.Fe(CO)_2]^-$	7×10^7	$[Re(CO)_5]^-$	2.5×10^4
$[\eta\text{-}C_5H_5.Ru(CO)_2]^-$	7.5×10^6	$[Mn(CO)_5]^-$	77

$[\eta\text{-}C_5H_5 . W(CO)_3]^-$ 500 $[Co(CO)_4]^-$ 1

$[\eta\text{-}C_5H_5 . Mo(CO)_3]^-$ 67

$[\eta\text{-}C_5H_5 . Cr(CO)_3]^-$ 4

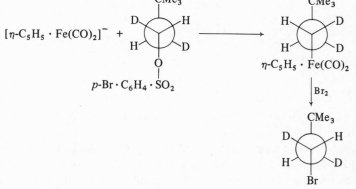

It was initially believed that one of the factors that determined the relative nucleophilicities of complexes was the co-ordination number of the metal. For metal carbonyls and their anions the most stable co-ordination numbers are 4 and 6, with 5 being less favourable. The increase in the co-ordination number from 5 to 6 when a metal–carbon bond is formed as the result of nucleophilic attack by $[Mn(CO)_5]^-$ was therefore regarded as more favourable than the change from 4 to 5 associated with the corresponding reaction of $[Co(CO)_4]^-$. However, a comparison of the nucleophilicities of an extensive range of metal complexes reveals that there is no obvious way of correlating the nucleophilicities of metals in different groups of the Periodic Table (as with manganese and cobalt). Indeed, even with metals in the same group, nucleophilicities appear to depend upon oxidation states. Thus while the order Rh > Ir > Co is observed with the anionic complexes of type $[M(CO)_2\text{-}(PPh_3)_2]^-$, this order is completely reversed with certain neutral complexes. One trend that has been repeatedly observed, however, is that as with basicity, the nucleophilicity of a particular metal is dependent upon the nature of the ligands, and with anionic carbonyls the replacement of a co-ordinated carbon monoxide by another ligand that is a stronger σ-donor but weaker π-acceptor invariably increases the nucleophilicity on account of the increased electron density on the metal:

$$[Co(CO)_2 [P(OPh)_3]_2]^- > [Co(CO)_3 [P(OPh)_3]]^- > [Co(CO)_4]^-$$

Nucleophilic substitutions with metal complexes as the nucleophiles also occur at unsaturated carbon, and with acyl and aroyl halides all the anionic

complexes mentioned so far in this discussion afford the expected acyl–
and aroyl–metal derivatives.

$$[\eta\text{-}C_5H_5 \cdot Fe(CO)_2]^- \xrightarrow{Ph \cdot CO \cdot Cl} \eta\text{-}C_5H_5 \cdot Fe(CO)_2 \cdot CO \cdot Ph$$

$$[Mn(CO)_5]^- \xrightarrow{Me \cdot CO \cdot Cl} Me \cdot CO \cdot Mn(CO)_5$$

In most cases these derivatives can be thermally decarbonylated to give
the corresponding alkyl and aryl compounds, but the ease with which this
can be accomplished varies greatly. This point is conveniently illustrated by
the decarbonylation of the perfluoropropyl derivatives (LXXVI). While
(LXXVIa) decarbonylates spontaneously at room temperature, (LXXVIb)
and (LXXVIc) require several hours heating at 80 °C and 120 °C respectively.
The iron compound (LXXVId) is thermally stable and has to be irradiated
with ultraviolet light for the decarbonylation to take place.

$$C_3F_7 \cdot CO \cdot M \xrightarrow{\Delta} C_3F_7 \cdot M + CO$$

(LXXVI) a $M = Co(CO)_4$
          b $M = Mn(CO)_5$
          c $M = Re(CO)_5$
          d $M = \eta\text{-}C_5H_5 \cdot Fe(CO)_2$

Isotopic labelling experiments have shown that in the decarbonylation of
acyl metal carbonyls the carbon monoxide that is evolved is one of the
CO ligands originally bonded to the metal, and not that of the acyl group.

$$Me \cdot \overset{*}{C}O \cdot Mn(CO)_5 \xrightarrow{\Delta} Me \cdot Mn(\overset{*}{C}O)(CO)_4 + CO$$

The significance of this result and other aspects of decarbonylation are
discussed on p. 270.

The more nucleophilic complexes can also participate in nucleophilic
substitutions with aryl halides. Indeed, the species $[\eta\text{-}C_5H_5 . Fe(CO)_2]^-$
even reacts at room temperature with iodobenzene to give the phenyl derivative $\eta\text{-}C_5H_5 . Fe(CO)_2 . Ph$, although only a very low yield (2 per cent) is
obtained after a reaction time of four days. The less reactive nucleophiles
require fairly reactive substrates such as perfluorobenzene or its heterocyclic
analogues.

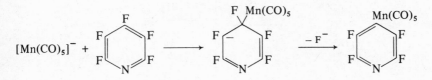

With these fluorinated aromatics the nucleophilic substitutions are promoted because the fluorine atoms can stabilise the negative charge introduced into the system by the attacking nucleophile. This is also true of those perfluoro-alkenes which undergo nucleophilic substitutions with metal complexes.

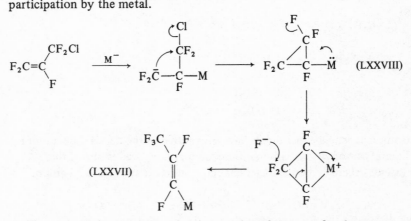

The formation of a carbanion in these substitutions at unsaturated carbon occasionally leads to the formation of 'anomalous' products. Thus several anionic metal carbonyls such as $[\eta\text{-}C_5H_5 \cdot Fe(CO)_2]^-$ and $[Re(CO)_5]^-$ react with perfluoroallyl chloride to give *trans*-perfluoropropenyl derivatives (LXXVII). This reaction can be visualised as involving a metal–cyclopropyl intermediate (LXXVIII), the ring of which readily opens as the result of participation by the metal.

This proposed participation by the metal is of interest, for there are a number of reactions of transition-metal complexes which can also be rationalised in terms of the metal helping to stabilise a positive charge that is generated at the β-position of a σ-bonded ligand (see useful discussion by Coates, Green and Wade, 1968, and many examples in the review by Rosenblum, 1974). Examples of such reactions include hydride abstraction from an alkyl group and protonation of a β-keto group, both of which can be illustrated by reference to the cyclopentadienyldicarbonyliron system.

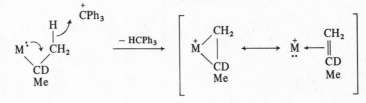

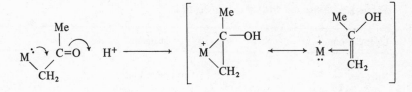

$$M = \eta\text{-}C_5H_5 \cdot Fe(CO)_2$$

The very high stability observed for certain types of carbonium ions such as (LXXIX), (LXXX) and (LXXXI), can similarly be explained by assuming that the empty p-orbital on the positively-charged carbon atom overlaps with a filled 'non-bonding' d-orbital on the metal (see also p. 125).

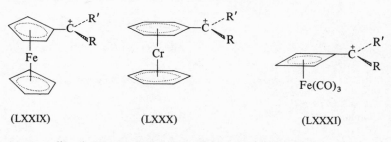

(LXXIX)   (LXXX)   (LXXXI)

Not unexpectedly, those metal complexes which undergo reactions in which the metal appears to stabilise a positive charge invariably have structures from which one would predict that there is a high electron density on the metal.

1.8.   σ-Donor/π-donor ligands

It was pointed out on p. 4 that a σ-donor ligand can also act as a π-donor if in addition to the filled orbital responsible for the σ-donor properties the ligand atom bears a filled or partly filled orbital which can participate in π-bonding by overlapping with an empty orbital on the metal. As with π-acceptor ligands, this π-bonding can be represented by the appropriate resonance structures, as shown by (LXXXII) and (LXXXIII) for F^- and HN^{2-} bonded to a bivalent metal.

$$\overset{+}{M}-\ddot{\underset{}{F}} \longleftrightarrow M=\overset{+}{\underset{}{F}} \qquad \text{(LXXXII)}$$

$$\overset{+}{M}-\bar{N}H \longleftrightarrow M=NH \qquad \text{(LXXXIII)}$$

The number of basic types of π-donor ligands is small compared with the number of π-acceptor ligands, and the more important ones are listed in table 1.6. The presence of a lone pair, a negative charge, or both of these

Table 1.6. *Some σ-donor/π-donor ligands*

Ligand atom		Type of ligand
Nitrogen	$\overline{}N\diagdown$	Amidate anions
	:N⟨ring⟩R	Pyridine-type heterocycles with electron-releasing substituents, e.g. R = OR′, NR′$_2$
	:N⟨ring⟩NH	Imidazole
	$\overline{}N=C=S$	Isothiocyanate anion
Oxygen	$^2\,\overline{}O$	Oxide anion
	$\overline{}O-$	Alkoxide and phenoxide anions
	$:O=C\diagup\diagdown$	Carboxylic acid derivatives, ureas
Sulphur	$:S=C\diagup\diagdown$	Thiocarbonyl compounds
	$:S=C\diagdown^{S^-}$	Dithiocarboxylate, xanthate and dithiocarbamate anions
Halogen	$\overline{}X$	Halide anions

indicates the ligand atom or atoms. These are atoms of nitrogen, oxygen, sulphur, or a halogen.

In π-donor ligands where the ligand atom forms part of a conjugated system, this atom bears a partial negative charge as the result of resonance stabilisation. This is indicated by the zwitterion type of resonance structures which can be drawn for such ligands, e.g. imidazole.

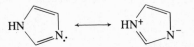

As a general rule, the π-donor properties of these ligands increase as the atom in the zwitterion structure on which the positive charge is placed is changed from sulphur to oxygen to nitrogen; e.g. with carboxylic acid derivatives (LXXXIV), where the carbonyl oxygen is the ligand atom, π-donor strength increases in the order thiol esters < esters < amides.

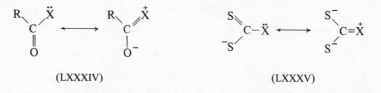

(LXXXIV)   (LXXXV)

Similarly, dithiocarbamates (LXXXV; X = NR$_2$) are stronger (chelating) π-donors than xanthates (LXXXV; X = OR).

Some π-donor ligands are very effective at stabilising metals in high oxidation states, and they therefore contrast with π-acceptor ligands which generally stabilise low oxidation states. Until recently the highest oxidation states of the transition metals were almost always exhibited in their oxides and fluorides, and in the case of oxides one reason for this is the ability of the O^{2-} ligand to form strong π-bonds with highly charged metal cations. However, certain organic ligands such as *N,N*-dialkyl dithiocarbamates (LXXXV; X = NR$_2$) have now been found to stabilise cations such as Ni^{4+}, Fe^{4+} and Cu^{3+}.

It should be noted, however, that the presence of π-donor ligands is not a necessary requirement for the stabilisation of high oxidation states, for very strong σ-donors that cannot participate in π-bonding, e.g. (LXXXVI), can also stabilise Ni^{3+}, Cu^{3+}, Ag^{3+} and other highly oxidised cations.

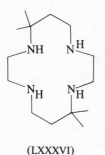

(LXXXVI)

One interesting result of the ability of the halide anions to π-bond with a metal is that with Group IIIB elements (notably boron and aluminium) the Lewis acidities of the halides increase in the unexpected order fluoride < chloride < bromide < iodide. This topic is discussed on p. 59.

1.9. Metal–ligand affinities; Lewis acidity and basicity

Many reactions in which metal complexes participate include an early stage where one or more of the reactants co-ordinates with the metal. Numerous examples of such reactions are discussed in chapters 2–5 but an obvious one to illustrate the point is Friedel–Crafts alkylation, which is initiated by the alkyl halide co-ordinating with the catalyst.

$$Al_2Cl_6 + 2R \cdot Cl \longrightarrow 2 \quad \begin{array}{c} Cl \diagdown \quad \diagup Cl \\ Al \\ Cl \diagup \quad \diagup^{\uparrow} \quad Cl \cdot R \end{array}$$

In the majority of reactions this co-ordination involves the reactant displacing a ligand already present on the metal, but in some reactions a vacant co-ordination position is already available. Regardless of the actual mechanism of co-ordination however, it is clear that if co-ordination is an early step in the reaction pathway, the ease with which it occurs can be an important factor (although not necessarily a decisive one) as far as the overall rate of reaction is concerned. It is therefore worthwhile considering some of the factors that determine how readily ligands co-ordinate with metals, particularly those factors that are thermodynamic in origin and that determine the relative affinities different metals have for the various types of ligands. Before doing this in §1.10 an indication will be given of how the affinity between a specific metal and a specific ligand is defined and measured quantitatively.

A high proportion of the vast bulk of thermodynamic data now available on metal–ligand affinities have been obtained by studying the reversible, stepwise displacement of the co-ordinated water molecules from metal cations in aqueous solution, a process which can be represented by the following general series of equations.

$$M(H_2O)_n + L \quad \rightleftharpoons \quad M(H_2O)_{n-1} L + H_2O$$

$$M(H_2O)_{n-1} L + L \rightleftharpoons \quad M(H_2O)_{n-2} L_2 + H_2O$$

$$\cdots$$

$$M(H_2O)L_{n-1} + L \rightleftharpoons \quad ML_n + H_2O$$

In practice the number of co-ordinated water molecules initially present is often unknown, and they are frequently omitted from the equations for a specific system, as in the following example.

$$Ag^+ + NH_3 \rightleftharpoons Ag^+(NH_3)$$

$$Ag^+(NH_3) + NH_3 \rightleftharpoons Ag^+(NH_3)_2$$

As such displacements are reversible, equilibrium constants can be defined for every system, as in the case above.

$$K_1 = \frac{[Ag^+(NH_3)]}{[Ag^+][NH_3]} \quad K_2 = \frac{[Ag^+(NH_3)_2]}{[Ag^+(NH_3)][NH_3]}$$

These constants are termed *stepwise stability constants* since they reflect the stability of the various species formed by stepwise displacement of the co-ordinated water molecules, and they may be measured in a number of ways (Rossotti and Rossotti, 1961). For statistical, electrostatic and steric reasons, the values of these constants nearly always decrease in the order $K_1 > K_2 > K_3 > \ldots > K_{n-1} > K_n$.

A more useful indicator of stability is provided by the so-called *overall stability constants*, for which the symbol β is used, as in:

$$\beta_1 = \frac{[Ag^+(NH_3)]}{[Ag^+][NH_3]} \qquad \beta_2 = \frac{[Ag^+(NH_3)_2]}{[Ag^+][NH_3]^2}$$

The general equation for β_n is as follows.

$$\beta_n = \frac{[ML_n]}{[M][L]^n} = K_1 \times K_2 \times K_3 \times \ldots \times K_n$$

Values of overall stability constants cover a very wide range, and can be less than unity for a comparatively unstable species but as high as 10^{40} for an exceptionally stable one. For this reason the values of these constants, and also those of stepwise stability constants, are usually quoted on a logarithmic scale:

$$pK = \log_{10} K \qquad p^\beta = \log_{10} \beta$$

As a very rough guide, a species that would normally be considered as 'stable' would have a $p\beta$ value of greater than 7 or 8, but it should be noted that this is a consideration of stability only in the thermodynamic sense of the word, and is in no way related to the stability of the species with respect to oxidation, hydrolysis, thermal degradation, or any of the other processes by which co-ordination compounds may be destroyed.

The values of the overall stability constants associated with complex formation between the nitrogen ligands, ammonia and pyridine, and Cu^{2+} and Ni^{2+} are given in table 1.7. These values are taken from the extensive collection of metal–ligand stability constants compiled by J. Bjerrum, Schwarzenbach and Sillen (1957, 1958) and Sillen and Martell (1964, 1971).

Table 1.7. *Stepwise and overall stability constants measured at 25 °C in aqueous solution*

System	pK_1	pK_2	pK_3	pK_4	$p\beta_4$
Ni^{2+}/NH_3 [a]	2.4	1.9	1.6	1.2	7.1
Cu^{2+}/NH_3 [a]	4.3	3.6	2.9	2.2	13.0
Ni^{2+}/py [b]	2.1	1.7	1.1	0.6	5.5
Cu^{2+}/py [b]	2.5	2.0	1.3	0.8	6.6

[a] $[NH_4NO_3] = 1M.$ [b] $[NaClO_4] = 1M.$

An inspection of the stability constants in the table reveals several points including the following:

(1) Both metals have a higher affinity for the two nitrogen ligands than for water ($pK_1 > 0$).

(2) Both metals have a higher affinity for ammonia than for pyridine.

(3) Cu^{2+} has a higher affinity for both nitrogen ligands than has Ni^{2+}.

On the basis that the two nitrogen ligands and water are Lewis bases (electron donors) while the two metal cations are Lewis acids (electron acceptors), and that the extent to which a Lewis acid and base enter into complex formation is a measure of their Lewis acidity and basicity, we may conclude that with reference to both the Lewis acids Cu^{2+} and Ni^{2+} the Lewis basicity of the three ligands increases in the order $H_2O < py < NH_3$, and that with reference to all three ligands the Lewis acidity of Cu^{2+} is higher than that of Ni^{2+}. By similarly comparing other appropriate sets of stability constants it is possible to draw up orders of Lewis acidity and Lewis basicity for an extensive range of metal ions and ligands respectively. One then finds, for example, that with respect to a very large number of ligands the Lewis acidity of First Row dipositive transition-metal ions changes in the order:

$$Mn < Fe < Co < Ni < Cu > Zn$$

This is called the Irving–Williams order. In spite of this specific general observation it must be emphasised that whenever an order of Lewis acidity is being assigned to a collection of metal cations the reference ligand must always be defined, for a different reference ligand can often lead to a different order of Lewis acidity. The same observation, of course, also applies to orders of Lewis basicity.

Although this technique of assigning relative Lewis acidities and basicities appears at first sight to be of universal application, it has in fact several serious limitations. One of these is that the technique is applicable only to *kinetically labile* cations, i.e. cations such as those listed in the Irving-Williams order whose ligand substitution reactions reach their equilibrium position fairly rapidly. In practice this means well within one minute at room temperature for 0.1M solutions. The technique is not applicable to certain cations such as Cr^{3+}, Co^{3+} and Rh^{3+} whose ligand substitution reactions take place extremely slowly and therefore require considerable periods of time to reach equilibrium. Such cations are said to be *kinetically inert* (see review by Taube, 1952).

Another disadvantage of determining relative Lewis acidities and basicities as described above is that the method is inapplicable to a number of Lewis acids such as BF_3, $AlCl_3$, $ZnCl_2$ and $SnCl_4$ which are of considerable importance because of their use in preparative organic chemistry. These compounds are not only extremely sensitive to water, but they are able to interact with Lewis bases without undergoing ligand substitution because vacant coordination positions are already available on the central atom.

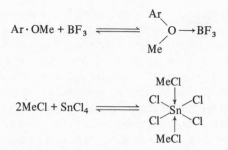

Strictly speaking, compounds such as the halides of aluminium and zinc should not be included in this category, for these compounds are oligomeric or polymeric in nature, with the co-ordination positions which are subsequently occupied by the donor atom of the Lewis base being filled by means of halogen bridges.

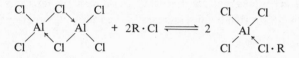

Indeed, it can be argued that even with compounds such as boron trifluoride, complex formation is usually performed in a solvent, and under these conditions the 'vacant' co-ordination position is already occupied by a solvent molecule which is subsequently displaced.

With these formally co-ordinatively unsaturated Lewis acids it is occasionally possible to make a qualitative estimation of the relative Lewis basicities of various ligands by means of displacement studies similar to those used for obtaining stepwise and overall stability constants in aqueous media. Thus, the fact that ethyl acetate can completely displace diethyl ether from the complex $Et_2O . AlCl_3$ suggests that, with reference to aluminium trichloride, the ester is the stronger Lewis base of the two ligands. A more general procedure, however, is to compare enthalpies of complex formation. If a specific Lewis acid (M) combines separately and reversibly with a number of ligands, the relative Lewis basicities of these ligands are indicated by the values of the equilibrium constant (K) for the equilibria of the general type:

$$M + L \underset{}{\overset{K}{\rightleftharpoons}} ML$$

Now the equilibrium constant is related to the enthalpy change (ΔH^0) and other thermodynamic functions by the equation:

$$-RT \ln K = \Delta G^0 = \Delta H^0 - T \Delta S^0$$

If there are good reasons for believing that the entropy change (ΔS^0) is similar for all the equilibria, then the relative Lewis basicities of the various ligands can obviously be deduced from the value of the enthalpy change. This can be determined for each ligand by direct calorimetric measurement or graphically by plotting the stability constant against temperature. The former method is superior not only for the systems under discussion but also for those generated in aqueous media. Ideally, the enthalpy changes considered should be those of complex formation in the gas phase in order to ensure the absence of enthalpy contributions from those factors such as solvation and lattice energies which are invariably associated with condensed phases. In practice, non-polar non-co-ordinating solvents can often be used as the medium, with carbon tetrachloride being preferred for oxygen donors, and cyclohexane for nitrogen and sulphur ones.

This technique of using enthalpy changes for assessing the relative basicities of various Lewis bases with respect to a specific Lewis acid can obviously also be applied for assessing Lewis acidities with respect to a specific Lewis base. Thus, enthalpy values indicate that when pyridine is the reference base, Lewis acidities of Group IIIB halides increase in the following order:

$$AlX_3 > BX_3 > GaX_3$$

It is frequently assumed that in addition to describing the extent of complex formation, the thermodynamic functions ΔG^0 and ΔH^0 also give a direct measure of the *electron acceptor* (or donor) *strength* of a Lewis acid (or base). This is not necessarily so, for although ΔH^0 is often dominated by the donor–acceptor bond strength there are other enthalpy contributions associated with changes which take place in bonds adjacent to the site of reaction. These contributions may vary from one system to another. For example, when boron acts as the acceptor atom in a Lewis acid its hybridisation state changes from sp^2 to sp^3, and the planar trigonal Lewis acid is converted into a tetrahedral adduct:

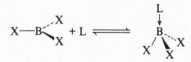

The enthalpy change associated with this reorganisation depends on the nature of the groups attached to the boron, and is more positive for BF_3 than for BH_3 (see also p. 59). As a result, although the trifluoride is by far the stronger electron acceptor, the overall ΔH^0 values found for complex formation are often approximately the same for both boron compounds.

In recent years this inability to directly obtain specific information on electron donor and electron acceptor strengths from ΔH^0 and ΔG^0 values has been overcome to a limited extent by means of spectroscopic techniques. One such technique uses the complex $\eta\text{-}C_5H_5 . Fe(CO)_2 . CN$ as the reference Lewis base. Co-ordination of the nitrogen lone pair on the cyano group as shown in (LXXXVII) makes this group a weaker σ-donor but a better π-acceptor (for the iron, that is). Both modifications reduce the electron density on the iron, and hence reduce the extent of back-bonding to the carbon monoxide ligands,

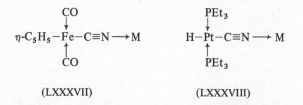

(LXXXVII)   (LXXXVIII)

i.e. ν_{CO} is increased (see p. 21). The greater the extent to which the nitrogen lone pair is involved in bond formation with the Lewis acid (M), the greater the increase in ν_{CO}. Similarly the coupling constant J(Pt–H) for complexes of type (LXXXVIII) reflects the acceptor strength of the Lewis acid M, which is shown by this technique to decrease as one passes along the series M = $AlCl_3$, $CoCl_2$, $ZnCl_2$, $AlMe_3$.

One final point concerning Lewis acidity is that it is unwise to assume that there is a direct correlation between the strength of a Lewis acid and the efficiency with which it can promote or catalyse reactions. This efficiency is usually determined by a number of factors other than the strength of the acid, and whose relative importance can vary from system to system. Consequently a Lewis acid that may be very efficient at promoting one type of reaction may be completely ineffective elsewhere.

1.10   Metal–ligand affinities; electronic effects

Having seen in §1.9 how the extent of complex formation between Lewis acids and Lewis bases can be measured, it is appropriate to extend the discussion to the results of such measurements, and – more specifically – to consider the factors that determine the relative stabilities of metal complexes and the relative affinities different metals have for the various ligands.

One generalisation that can be made (Ahrland, Chatt and Davies, 1958) is that metals tend to fall into two classes, the so-called classes (*a*) and (*b*). Class (a) metals include the alkali metals, alkaline earth metals and the lighter transition metals in higher oxidation states (e.g. Co^{3+}, Ti^{4+} and Fe^{3+}). The stability of complexes formed between metals in this class and structurally related ligands varies with the ligand atom in the following orders:

$$N \gg P > As > Sb$$

$$O \gg S > Se > Te$$

$$F > Cl > Br > I$$

Class (*b*) metals are mainly the heavier transition metals (e.g. Hg^{2+}, Pt^{2+}) and the lighter ones in low oxidation states (e.g. Cu^+), and the stability of their complexes varies with the ligand atom in the following orders:

$$N \ll P > As > Sb$$

$$O \ll S < Se \sim Te$$

$$F < Cl < Br < I$$

Thus while the class (*a*) metals Mg^{2+}, Al^{3+} and Fe^{3+} prefer to bond with nitrogen and oxygen (e.g. ammonia and water), the class (*b*) ones Hg^{2+} and Pd^{2+} prefer phosphorus and sulphur (e.g. phosphines and thioethers).

According to their preference for either class (*a*) or class (*b*) metals, ligands may be classified as type (*a*) or (*b*) respectively.

These generalisations have been extended by Pearson (1963) into a more general principle which covers many types of Lewis acids and bases: hard acids prefer to bond to hard bases, and soft acids prefer to bond to soft bases. 'Hard' acids and bases are generally small in size and not readily polarised, while 'soft' acids and bases are generally large and easily polarised. Examples of the four different types are as follows:

Hard acids H^+, Na^+, Mg^{2+}, Ti^{4+}, Fe^{3+}, Al^{3+}, CO_2, SO_3.
Soft acids Pd^{2+}, Ag^+, Hg^+, $:CH_2$, Br_2, 1,3 5-trinitrobenzene.
Hard bases NH_3, H_2O, AcO^-, F^-.
Soft bases C_2H_4, CO, R_3P, R_2S, I^-.

On Pearson's classification, class (*a*) metals and ligands are hard, while class (*b*) metals and ligands are soft. Most class (*b*) metals are able to participate in π-bonding, and this may be one of the reasons for their preference for sulphur and phosphorus, which are π-acceptor ligand atoms (empty 3d-orbitals), rather than oxygen and nitrogen. In contrast, class (*a*) metals are usually linked to their ligands predominantly by σ-bonds which have a relatively high degree of ionic character. Generally speaking, therefore, one finds that with a specific metal in this class there is a good linear correlation between the stability of the complexes formed with a series of closely related ligands and the basicity (i.e. affinity for a proton) of the ligands. The basicity of the bidentate anions (LXXXIX; *a–d*), for example, increases in alphabetical order, $a < b < c < d$, i.e. the *acidity* of the parent phenols increases in the

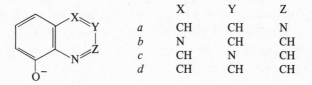

	X	Y	Z
a	CH	CH	N
b	N	CH	CH
c	CH	N	CH
d	CH	CH	CH

(LXXXIX)

reverse order, and a plot of basicity against the stability constant (pK_1) of the complexes formed with Mg^{2+}, Zn^{2+}, and Cu^{2+} is a straight line for each metal.

For the same reason, with a specific ligand the relative affinities of class (*a*) metals tend to be governed largely by ionic potential, i.e. the effective charge to radius ratio. Consequently, when a ligand forms complexes with a class (*a*) metal in more than one oxidation state those formed from the higher oxidation states are usually the more stable ($Fe^{3+} > Fe^{2+}$, $Co^{3+} > Co^{2+}$). The stabilities of complexes of the alkali metals decrease with increasing atomic radius of the metal, while those formed by the lanthanides increase steadily as one passes from lanthanum to lutecium, owing to a steady increase in ionic potential because of the lanthanide contraction. The ionic potential effect is also largely responsible for the order of stability observed for complexes of certain divalent metal cations with a very large number (>100) of ligands – mainly those which bond through oxygen and/or nitrogen:

$$Ba^{2+} < Sr^{2+} < Ca^{2+} < Mg^{2+} < Mn^{2+} < Fe^{2+} < Co^{2+} < Ni^{2+} < Cu^{2+} > Zn^{2+}$$

This order, which includes the Irving–Williams order for transition-metal cations, is particularly significant in that the ions listed can catalyse a number of organic reactions such as the hydrolysis of α-amino esters and the decarboxylation of certain keto acids. In most of these reactions catalysis occurs because the metal cation co-ordinates with one of the reactants by means of a ligand atom on which negative charge develops during the rate-determining step of the reaction. This negative charge is effectively stabilised by the metal, and the activation energy of the reaction is therefore lowered. In the catalysed hydrolysis of α-amino esters, for example, one of the reactive species is the chelated ester (XC), and when the carbonyl group is attacked by water in the rate-determining step negative charge builds up on the co-ordinated oxygen atom (see also p. 161):

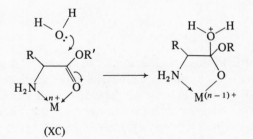

(XC)

Because both the ability of the metal cation to stabilise this negative charge and also the concentration of the reactive species are largely determined by the ionic potential of the metal, the order observed for the efficiency with which certain divalent metals catalyse the hydrolysis is consistent with their position in the sequence Ba^{2+} to Zn^{2+} given above.

As described in §1.9, most metal–ligand affinities are measured in aqueous solution, and under these conditions complex formation involves the displacement of one or more water molecules from the fully hydrated metal cation. Of greater interest to the organic chemist, however, is the affinity of the metal (for additional ligands) in Lewis acids that are formally co-ordinatively unsaturated, such as Friedel–Crafts catalysts, metalloporphyrins, and the substituted metal acetylacetonates used as shift reagents in n.m.r. spectroscopy. This affinity varies considerably from one metal to another even within the same group of the Periodic Table (cf. Ni and Pd, Zn and Hg below), and with a specific metal (main group or transition) it is dependent upon the ligands already present on the metal. Generally speaking, the affinity is increased by ligands that lower the electron density on the metal, e.g. weak σ-donor/strong π-acceptors, but decreased by those which increase this variable, e.g. strong σ-donors and moderate σ-donor/strong π-donors. This generalisation, which is illustrated in qualitative terms in the selected examples below, only applies of course when the ligand to be added is a relatively strong σ-donor and has little or no π-acceptor capacity, i.e. it is of the class (a) type. As described in the appropriate sections of this book, the affinity of a transition metal for weak σ-donor/strong π-acceptor ligands such as carbon monoxide, tetra-cyanoethylene and other unsaturated ligands which undergo oxidative addition, is decreased by ligands that lower the electron density on the metal. This is because with weak σ-donor/strong π-acceptor ligands the main contributor to the total metal–ligand bond strength is the metal–ligand π-bond strength, and this is lowered by any factor which decreases the degree of back-bonding to the ligand.

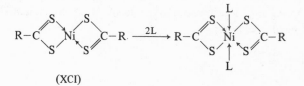

(XCI)

The high tendency for Lewis bases such as amines and ethers to occupy the fifth and sixth co-ordination positions of the metal in the xanthate complexes (XCI; R = O-alkyl) is inhibited when the σ-donor and π-donor properties of the ligands are increased. It is therefore lower in the *N*-aryl dithiocarbamate complexes (XCI; R = NH . aryl), and lower still in the *N,N*-dialkyl dithiocarbamates (XCI; R = N(alkyl)$_2$) where the resonance stabilisation of the $S = C—\ddot{N}R_2$ system is fully developed.

While Zn^{2+} is normally four co-ordinate and readily increases its co-ordination number to five and six by complex formation, the co-ordination number of Hg^{2+} is usually two, and this is increased (to three or four) much less readily than with Zn^{2+}. With the heavier metal, however, the increase is promoted by weak σ-donor ligands, and thus the perfluoro derivatives of bialkyl- and biaryl-mercury(II) compounds add extra ligands far more readily than the parent compounds. With $(NO_3)_3C \cdot Hg \cdot C(NO_3)_3$, complex formation occurs even with comparatively weak Lewis bases such as tetrahydrofuran.

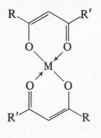

(XCII)

With the exceptions of palladium and platinum, the bivalent metals in acetylacetonates (XCII; $R = R' = Me$) accept additional ligands at the fifth (and in some cases such as nickel, also at the sixth) co-ordination position. Progressive replacement of the methyl hydrogens by fluorine steadily decreases the σ-donor and π-donor properties of the acetylacetonate ligands and also increases the ease with which the additional ligands are added. Equilibrium constants for the addition therefore increase in the order $(R = R' = CH_3) < (R = CH_3, R' = CF_3) < (R = R' = CF_3)$.

This effect of fluorination on the Lewis acidity of metal complexes has been used in the development of more effective n.m.r. shift reagents. These reagents are chelates, usually of β-diketones and paramagnetic lanthanides such as Eu^{3+}, Pr^{3+} and Yb^{3+}, which are added to a solution of the substrate under investigation. A donor atom in the substrate (usually oxygen or nitrogen as the lanthanides are class (a) metals) co-ordinates with the metal cation, and the chemical shifts of nuclei (^1H, ^{13}C, ^{19}F) in the vicinity of the donor atom are altered, largely as the result of dipolar inter-actions between these nuclei and the electron spin magnetisation of the paramagnetic metal (pseudocontact shift). As the exchange between co-ordinated substrate and unco-ordinated substrate is rapid on the n.m.r. timescale, a time-averaged spectrum is recorded, with the observed chemical shifts being dependent upon factors such as the stereochemical environment of the nuclei, the distance between the nuclei and the metal, and – most important in the present context – the extent of complex formation between the reagent and the substrate. The direction in which the chemical shifts change depends upon the metal, and are upfield for Pr^{3+} but downfield for Eu^{3+}. The overall effect is an increase in spectral resolution, and in many cases the production of a first-order spectrum. Because they are stronger Lewis acids, shift reagents with fully or partially fluorinated β-diketonate residues enter into complex formation to a greater extent than the unsubstituted ligands and therefore result in larger chemical shifts. Solubility is also an important factor, of course, and this is enhanced by chain branching which inhibits intermolecular metal–ligand interactions in the solid state. Both features are manifested by the ligand (XCIII), whose europium derivative, Eu(FOD)$_3$, is extensively used.

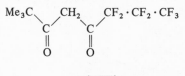

(XCIII)

One particular application of shift reagents is the determination of enantiomeric purity. When the n.m.r. spectra of enantiomers are recorded in the presence of chiral shift reagents such as (XCIV; *a–c*) the chemical shifts of the enantiotopic nuclei are usually quite different. Consequently the proportions of enantiomers in a mixture can be determined directly from its n.m.r. spectrum by integration of the appropriate signals. This technique provides a very convenient alternative to the time-consuming classical methods.

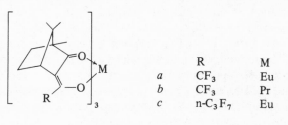

	R	M
a	CF$_3$	Eu
b	CF$_3$	Pr
c	n-C$_3$F$_7$	Eu

(XCIV)

Shift reagents and their applications have been reviewed by Mayo (1973) and Cockerill *et al*. (1973).

Fluorination of a ligand represents a comparatively simple electronic modification in that the three main possible effects, i.e. a reduction in the σ-donor strength, a reduction in the π-donor strength, and an increase in the π-acceptor strength of the ligand, all operate in the same direction and tend to reduce the electron density on the metal. The overall effect on the affinity of the metal for an additional ligand is therefore generally quite predictable, provided that one has a knowledge of the relative size of the σ-donor strength and π-acceptor strength (if any) of the ligand to be added. Thus, while fluorination of a metal's ligands increases the affinity of the metal for σ-donors with little or no π-acceptor capacity (t-amines, ethers), the affinity for weak σ-donor/strong π-acceptor ligands (carbon monoxide, tetracyanoethene) is decreased. The situation is not so straightforward, however, when a modification (or replacement) of a ligand results in changes in the σ-donor, π-donor

and π-acceptor strengths which counteract each other, e.g. the σ-donor
strength is decreased but the π-donor strength is increased. In this case the
affinity of the metal for an additional ligand is also determined by the relative
importance of the σ- and π-bonding between the metal and the ligand being
modified (or replaced). This is illustrated by halides of the Group IIIB metals,
boron,* aluminium and gallium, which can act as Lewis acids because there
is an empty p-orbital on the metal. With a particular metal the Lewis acidity
of these halides generally increases in the order fluoride < chloride < bromide
< iodide, i.e. the exact reverse of the order expected on the basis of the
relative σ-donor strengths of the halide anions. The main reason for this
anomaly is that in these metal halides the metal–halogen bonds contain a
π-component which is formed by overlap of a filled p-orbital on the halogen
with the empty p-orbital on the metal. As the latter orbital is fully utilised in
σ-bond formation when the metal halide co-ordinates with a Lewis base, this
π-component is completely destroyed by complex formation. Now it appears
that the strength of the π-component increases in the order iodide < bromide
< chloride < fluoride, i.e. the amount of π-bond energy that is lost on complex
formation increases as the atomic weight of the halogen decreases. Evidently,
as far as the extent of complex formation is concerned, this is a more important factor than the corresponding decrease in the σ-donor strength of the
halogen.

1.11.   Chelation and macrocyclic ligands

In the preceding section metal–ligand affinities were discussed solely in terms
of electronic effects. Stereochemistry of the ligand can also play an important
role, however, as dramatically illustrated by those multidentate ligands whose
stereochemistry allows two or more donor atoms to be simultaneously bonded
to the same metal. In general, these ligands form complexes (chelates) which
have relatively high stability constants, i.e. in the presence of a metal the
concentration of free ligand in solution is very low. This fact is utilised in
organic synthesis when a reaction product is trapped by chelating it with a
metal, thus ensuring that the rate at which the product undergoes any further
reaction is considerably reduced. Examples of this are provided by cross-aldol
condensations carried out with preformed metal enolates and in the presence
of halides of magnesium or zinc, and by the Grignard synthesis of mixed
ketones from esters of 2-pyridinethiol; both these reactions give excellent
yields.

---

* Boron is normally classified as a metalloid, but it is considered in this book as
  its complexes are analogous to those formed by metals and are frequently
  encountered in organic chemistry. It is referred to as a metal solely in order
  to simplify the text.

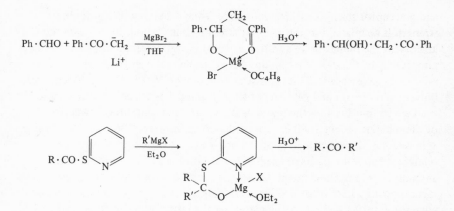

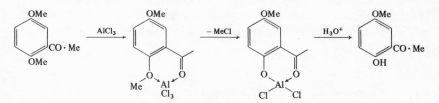

Functional groups which have only weak co-ordinating properties can be encouraged (or even forced) to co-ordinate with a metal by making them part of a chelating system. This is the basis of a number of selective reactions, such as the monodemethylation of aromatic ethers (cf. p. 134).

Many other examples of the use of chelation in organic chemistry can be found in later chapters of this book.

The metal complexes formed by chelating ligands are nearly always more stable than those formed by their monodentate analogues (the chelate effect) and this is illustrated quantitatively by the thermodynamic data obtained from the interaction of metal ions with the bidentate ligand 1,2-diamino-ethane (en) and with the monodentate one, methylamine. Complexes formed from the former ligand are invariably the more stable, and with Cd^{2+}, for example, a comparison of the appropriate stability constants shows that $[Cd(en)]^{2+}$ is about 10 times more stable than $[Cd(NH_2 . Me)_2]^{2+}$, while $[Cd(en)_2]^{2+}$ is about 10^4 times more stable than $[Cd(NH_2 . Me)_4]^{2+}$. An examination of the thermodynamic quantities show that with these two ligands the corresponding enthalpy changes for co-ordination to Cd^{2+} are almost identical, and that the greater stability of the diaminoethane complexes is due entirely to a more favourable entropy factor. In fact, entropy is invariably the factor which results in chelates being more stable than the corresponding complexes. More specifically, the entropy change $(T \Delta S)$ associated with chelate formation is substantially more positive than that associated with complex formation, and can often be large enough to compensate for an unfavourable enthalpy term.

It is generally accepted that the larger entropy increase associated with chelate formation is the direct result of a net increase in the total number of species in the system. Thus the displacement of the four co-ordinated water molecules from Cd^{2+} by two diaminoethane entities results in a net increase of two species, unlike the displacement by four ammonias which results in no increase. One point to notice, however, is that in certain respects chelation is an unfavourable process from the entropy viewpoint. When 1,2-diamino-ethane, for example, chelates with a metal, the free rotation about the carbon–carbon and carbon–nitrogen bonds which occurs in the unco-ordinated ligand is no longer possible, and consequently there is a corresponding loss of rotational entropy. Obviously this, together with the other unfavourable factors associated with the ligand being held in a restricted conformation, is not as important as the increase in translational entropy which results from an increase in the total number of species in the system.

The heterocyclic rings – or metallocycles, as they are termed – formed by chelation are similar to other organic ring systems in that their stability depends on size, with five- and six-membered rings being the most stable. The relative instability of medium sized chelate rings (8–11 members) is presumably due to strain which arises from factors such as non-bonded interactions, but this is a subject which has received little attention. There is increasing evidence, however, that as with the cycloalkanes, large or macrocyclic chelate rings (12 or more members) are relatively free of strain and no difficulty is experienced in preparing the chelates $IrCl(CO)[Bu_2^tP(CH_2)_{10}PBu_2^t]$ and $PtCl_2[Bu_2^tP(CH_2)_{12}PBu_2^t]$, for example, which contain thirteen- and fifteen-membered rings respectively.

With bidentate ligands, five-membered ring chelates are usually more stable than the analogous six-membered systems. Stability constants show, for example, that with metals such as Co^{2+}, Ni^{2+}, Cu^{2+}, Zn^{2+} and Cd^{2+}, the 1:1 chelates formed from 1,2-diaminoethane, glycine and oxalic acid are about 10–30 times more stable than those formed from corresponding homologues: 1,3-diaminopropane, 3-aminopropionic acid and malonic acid. The lower stability of the larger sized ring is partly due to an increase in the loss of entropy associated with ring formation, but is mainly due to a decrease in $-\Delta H^0$ because of greater ring strain. Thus while the $p\beta$ value of $Cu(gly)_2$ is 15.0, this value is 12.5 with the 3-aminopropionic acid analogue, and the decrease is almost entirely due to a decrease in $-\Delta H^0$ from 56 to 45 kJ mol^{-1}.

Interestingly, it appears that angle strain is also largely responsible for the fact that when a consecutive series of chelate rings is formed by a metal chelating with a multidentate ligand, the stability of the resultant chelate is often lowest when all the rings are of the same size – be they five- or six-

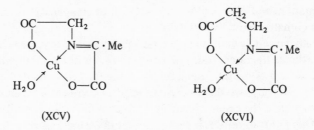

(XCV)   (XCVI)

membered – but is highest when the ring sizes alternate, e.g. 6,5,6, and 5,6,5. Thus the copper(II) chelate (XCVI) is more stable than (XCV), while the stability of the following amido complexes decreases from left to right.

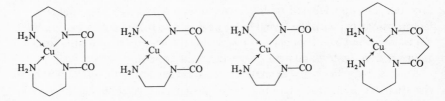

One notable exception to this generalisation is provided by ethylene-diaminetetra-acetic acid, a very strong chelating agent used in many branches of science and medicine on account of the high stability of its metal chelates (see for example, Seven and Johnson, 1960). With a number of metal ions this sexadentate ligand forms octahedral chelates which contain five five-membered rings, but which are more stable than those formed by the higher homologue based on 1,3-diaminopropane. The calcium chelate (XCVII; $n = 2$), for example, is more stable than (XCVII; $n = 3$) by a factor of about $10^{3.5}$.

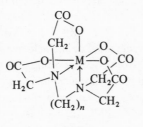

(XCVII)

The relative stabilities of various combinations of ring sizes reflect the ease with which the metal and ligand can mutually satisfy each other's stereochemical requirements. These requirements, of course, vary not only from ligand to ligand but also from metal to metal, and while Cu^{2+}, for example, requires a square-planar arrangement of its ligand atoms, four-co-ordinate Zn^{2+} requires a tetrahedral one. As a result, although the tetra-

dentate ligand (XCVIII) chelates with Cu^{2+} far more readily than the homologue (XCIX), both ligands chelate with Zn^{2+} with about the same ease. This
is revealed by the stability constants, which show the chelate formed by
(XCVIII) to be more stable than that formed by (XCIX) by a factor of about
6000 for Cu^{2+} but only by a factor of 5 for Zn^{2+}.

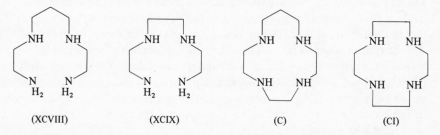

(XCVIII)   (XCIX)   (C)   (CI)

Related to these observations is the stabilisation of Cu^{2+} (with respect to
reduction to Cu^+) by chelating agents whose donor atoms can readily take up
the square-planar arrangement required by copper(II) but not the tetrahedral
one required by copper(I). Electrochemical reduction of the metal in copper(II)
chelates of porphyrins, for example, only occurs at very low negative potentials, and generally only after the ligand itself has undergone reduction.

As indicated earlier, although the overall entropy term for chelation is
usually more favourable than that for simple complex formation, this overall
term actually includes an unfavourable contribution which results from the
chelated ligand being unable to adopt numerous conformations that are
available to the unco-ordinated ligand. Now with certain macrocyclic chelating
ligands the ligand atoms are already held in approximately the correct positions
required by the metal, and the loss of conformational entropy associated with
chelate formation is therefore considerably less than that associated with
chelation by their acyclic analogues. As a result, these macrocyclic ligands are
the stronger chelating agents of the two types. The stability constants for the
chelation of Cu^{2+} by the macrocyclic tetramines (C) and (C1), for example,
are larger than those observed with the corresponding analogues (XCVIII)
and (XCIX) by factors of $10^{5.2}$ and $10^{4.8}$ respectively.

Another factor which can favour macrocyclic chelating ligands is the
reduced degree of solvation of the donor atoms which results from these
atoms being enclosed – and therefore sterically hindered to some extent – by
the ring system. This factor causes a favourable change in ΔH (heat of solvation of free ligand is lowered) and, as expected, is particularly important in
solvents such as water that can hydrogen-bond with ligand atoms. With
ligands where this factor is operative, the favourable entropy factor which
results from the ligand being macrocyclic tends to be offset by the fact that
fewer solvent molecules (lower increase in entropy) are displaced from the
ligand on chelation.

As with an acyclic chelating ligand, one factor that determines how strongly a macrocyclic ligand chelates with a specific metal is how readily both participants satisfy each other's stereochemical requirements. In this connection, the size of the metal and of the 'hole' in the middle of the macrocyclic ring are of obvious importance, for a ligand that can easily accommodate a small metal may experience considerable difficulty with a metal of much larger size. With porphyrins, for example, the size of the central 'hole' is such that in their metal derivatives a strainless, coplanar arrangement of the metal and the four nitrogen atoms requires that the ionic radius of the metal lie between the rather narrow limits of 60 and 69 pm, values that are exemplified by Ni^{2+} and Sn^{4+} respectively. This size requirement is one of the reasons why metalloporphyrins of Ag^{2+} (ionic radius, 84 pm) can be readily oxidised to the corresponding Ag^{3+} (ionic radius, 65 pm) systems, for the oxidations are associated with loss of radial strain.

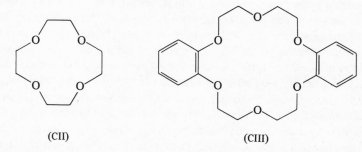

(CII)   (CIII)

The porphyrin nucleus is, of course, a fairly rigid and inflexible entity, but even macrocycles that have a high degree of flexibility can show a preference for specific metal ions which is based largely on ionic radius. The so-called 'crown' ethers are an excellent illustration of this. These are macrocyclic compounds in which the ring is built up from repeating $-CH_2.CH_2.O-$ units. As their formal names are rather long and cumbersome, they are usually given trivial names which indicate only the size of the macrocyclic ring and the number of oxygens in that ring. Thus (CII) is termed 12-crown-4. Simple derivatives, such as the frequently used dibenzo- and dicyclohexyl systems, e.g. (CIII), are easily named on the system. Crown ethers are of considerable interest in that they form complexes with alkali metal cations which are not only appreciably soluble in non-polar solvents, but which are more stable – both in solution and in the solid state – than those formed by other neutral ligands. The complex formed from K^+ and 18-crown-6, for example, has a stability constant of $10^{6.1}$ in methanol compared with $10^{2.2}$ for the potassium complex of the relatively strong chelating acyclic ligand $Me.(O.CH_2.CH_2)_5.OMe$, and is sufficiently soluble in benzene to enable salts such as potassium permanganate and potassium fluoride to dissolve in this solvent when the crown ether is present. In crown ether complexes the ligand oxygens are bonded to the metal purely by electrostatic forces and are located at discrete points on the co-ordination sphere. In some complexes such as that formed from (CIII) and potassium thiocyanate the gegen ion remains bonded to the metal, while in others the ring oxygens occupy all the available co-ordination positions, and the metal cation is then very efficiently separated from both the gegen ion and the solvent by the organic framework of the ligand. Under these conditions the basicity and nucleophilicity of the gegen ion are enormously enhanced, and consequently crown ethers may be used for promoting many reactions that involve 'weak' nucleophiles, e.g. substitutions with fluoride and carboxylate anions.

$$n\text{-}C_8H_{17}Br \xrightarrow[\text{18-crown-6}]{KF,\ MeCN} n\text{-}C_8H_{17}\cdot F$$

$$Ph\cdot CO\cdot CH_2Br \xrightarrow[\text{18-crown-6}]{R\cdot CO_2K,\ MeCN} Ph\cdot CO\cdot CH_2\cdot O\cdot CO\cdot R$$

The stability of the complexes formed from crown ethers and the alkali metals is highly dependent on the size of the macrocyclic ring and the radius of the cation, and is a maximum when the macrocycle can just accommodate the cation. Thus, the symmetrical dicyclohexyl derivatives of 12-crown-4, 18-crown-6 and 24-crown-8 form their most stable complexes with Na^+, K^+ and Cs^+ respectively. The crown ethers therefore contrast markedly with almost all acyclic ligands, which invariably form complexes whose stability decreases in the order $Na^+ > K^+ > Cs^+$. In this respect these macrocycles serve as model compounds for certain types of antibiotics which function as macrocyclic chelating agents by using ligand atoms in their ether, ester, or amide groups, and which selectively chelate with specific alkali-metal cations. Dicyclohexyl-18-crown-6, for example, is a useful model for nonactin (CIV), an antibiotic isolated from *Actinomyces* and which selectively enhances the transport of potassium ion through cell membranes by forming a stable,

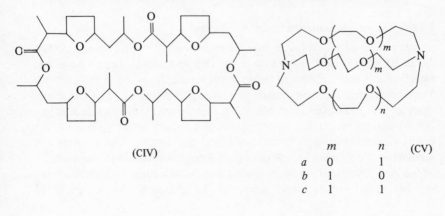

	m	n	
(CIV)			(CV)
a	0	1	
b	1	0	
c	1	1	

lipophilic chelate. In this chelate the ligand completely envelopes the potassium, which is bonded only to the four tetrahydrofuran oxygens and the four carbonyl oxygens, and is entirely devoid of solvent molecules.

Closely related to the crown ethers are those cage-like ligands whose metal complexes contain a metal cation encapsulated within an organic framework. For obvious dimensional reasons, the stability of these complexes or 'cryptates' (from the Latin, *crypta*, meaning a hole or cavity) is highly dependent upon the ionic radius of the metal, and thus with the alkali-metal cations the ligands (CV; a, b, and c) selectively complex with Li^+, Na^+, and K^+ respectively.

Many aspects of alkali-metal complexes of organic ligands (including crown ethers) have been fully reviewed by Dunitz *et al.* (1973), while the various applications of crown ethers to synthetic organic chemistry have been discussed by Gokel and Durst (1976).

Before leaving the subject of chelation it must be pointed out that although chelation stabilises complexes thermodynamically it also tends to stabilise them in other respects. Thus the thermal, oxidative, and hydrolytic stability of many types of organometallic systems is increased substantially if the metal is chelated, as in the following compounds.

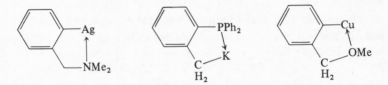

Similarly, the 1,2-diaminoethane chelate of ethynyl-lithium (Li . C≡CH) is a solid which is comparatively stable to hydrolysis by atmospheric moisture, and its use in the reaction with alkyl halides, for example, obviates the need for liquid ammonia as a solvent.

1.12. Steric hindrance in metal complexes

Another stereochemical factor that can be important in determining metal–ligand affinities is steric hindrance. In metal complexes this can arise within a particular ligand or between two or more ligands attached to the same metal. The concept that steric hindrance in a monodentate ligand can affect the affinity of that ligand for a metal was first introduced by Brown and Bartholomay (1944) who suggested that co-ordination of a tertiary amine with bulky substituents would create strain within the amine. This strain, termed B-strain on account of it occurring *behind* the ligand atom with respect to the metal, would arise because although in the unco-ordinated amine any steric interaction between the bulky substituents could be relieved to some extent by an increase in the C–N–C angles, this increase could not occur to the same extent in the co-ordinated amine which would strongly favour a regular tetrahedral arrangement of the bonds about the nitrogen atom. Co-ordination would therefore result in increased interaction between the substituents. Attractive as this concept is, there appears to be no convincing evidence that B-strain really exists, for up to now all the experimental evidence that can be used to support its possible existence can also be explained by alternative concepts.

For a more satisfying illustration of how internal strain within a ligand can affect its affinity for metals one must turn to chelating ligands. When such ligands are bonded to a metal they are often held in configurations or conformations that are sterically unfavourable (see §2.2), and this has an adverse effect on their chelating potential. Thus, chelates of the *meso*-form of 1,2-diamines of the type $H_2N.CHR.CHR.NH_2$ are less stable than those of the *dl*-form because in the former chelates the two R groups are on the same side of the five-membered chelate ring, and this leads to steric crowding which is absent in the latter chelates. In the copper(II) and nickel(II) systems the appropriate stability constants (K_1) differ by a factor of about $10^{0.7}$ for R = Me, and about 10^2 for R = Ph.

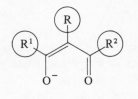

a	R = H, $R^1 = R^2 = CHMe_2$
b	R = H, $R^1 = R^2 = CMe_3$
c	R = $CHMe_2$, $R^1 = R^2 = Me$
d	R = CH(Me)Et, $R^1 = R^2 = Me$

(CVI)

In metal chelates the anion of a β-dicarbonyl compound is held in the configuration (CVI), which with the free ligand is unstable on account of the repulsion between the two negatively charged oxygens. The stability of chelates of this anion is highly dependent upon the substituents R, R^1 and R^2, and although it is often difficult to separate electronic effects it is evident that the size of these substituents is important. Steric interaction amongst the substituents is promoted by chain branching at the α-positions, and even with the anions (CVI; *a* and *b*) where one of the substituents is hydrogen the interaction appears to be sufficient to cause a reduction in the CO–C–CO bond angle and hence a reduction in the O–O distance, for these anions show exceptional chelating ability for the small Li^+ cation and may be used for effectively separating this cation from those of the other alkali metals. When R is an alkyl group the basicity of the anion is higher than when R is hydrogen, i.e. the parent dicarbonyl compound is a weaker acid, but the chelating ability is not as high as would be expected on this basis – indeed, in some cases it is actually lower than that of the non-alkylated anion. This has been attributed to steric hindrance in the chelates between the alkyl group and the terminal substituents R^1 and R^2, but an increase in the hydrophobic character of the ligand is also believed to be involved. With some systems branching at the α-position of the group (R) appears to suppress completely the chelating properties, for the β-diketones which would give the anions (CVI; c and d) on basification do not give the usual colour reactions indicative of chelate formation when treated with Cu^{2+} or Fe^{3+}.

Many clear-cut examples of steric hindrance between two or more ligands attached to the same metal are now known, but the bulk of the quantitative information on this subject concerns the stability of complexes of Group IIIB metals, notably boron, with monodentate amines, ethers and sulphides. The stability of these complexes is very sensitive to the size of the substituents attached to the ligand atom and to the metal, and in some systems the steric interaction between these substituents is sufficient to reverse the orders of stability expected on the basis of the usual Lewis basicities of the ligands. Thus although t-butylamine is both a stronger base and stronger σ-donor than ammonia, the complex formed from trimethylboron and the former amine is much less stable than that formed with the latter. Similarly, the stability of the 2-methylpyridine–trimethylboron complex is lower than that of the 3- and 4-methyl analogues, while the 2,6-dimethylpyridine complex is not formed at all, even at -80 °C. The boron trifluoride and boron trichloride complexes of diethyl ether are less stable than those of dimethyl ether, and the same order of stability is observed for the complexes of diethyl and dimethyl sulphides. As expected, the steric interaction is decreased as the distance between the metal and ligand atom is increased. Thus it is decreased when the metal atom is changed from boron to aluminium, and when the ligand atom is changed from oxygen to sulphur. It is largely eliminated, of course, if the substituents on the ligand atom are held back away from the metal, and thus very stable complexes are formed by the tertiary amine quinuclidine, while complexes of tetrahydrofuran and tetrahydropyran (and also the analogous sulphur heterocycles) are far more stable than those of their acyclic analogues.

Strain which arises as the result of this type of steric interaction has been termed F-strain by H. C. Brown – who has carried out extensive investigations in this area – on the grounds that the interactions responsible for the strain occurs mainly in *front* of the ligand atom with respect to the metal. For a critical but now rather dated review of H. C. Brown's work, see Hammond (1956).

In more recent years, this steric hindrance between large monodentate ligands has proved to be of synthetic use in the preparation of complexes in which the metal has an unusually low co-ordination number. Thus the use of bulky anions such as 2,6-dimethylpiperidide, $[NPr^i_2]^-$ and $[CH(SiMe_3)_2]^-$, has allowed a convenient preparation of neutral trigonal complexes of the type ML_3,

where M is Cr(III), Ti(III), or V(III). X-ray structural determinations show that
in these complexes the three bulky anions effectively shield the metal, and in
fact only ligands which have very small steric requirements, e.g. nitric oxide,
can be added. In solution, the colourless zerovalent nickel(0) complexes of
the type NiL_4, where L is a tertiary phosphine or phosphite, partly dissociate
to give the corresponding tris-complexes NiL_3, which are orange to red. The
degree of dissociation appears to be governed more by the size of the ligands
and their steric interaction than by electronic factors, and by using bulky
ligands such as PPh_3, $P(cyclohexyl)_3$ and $P(O-o-C_6H_4 . Me)_3$ appreciable
concentrations of the co-ordinatively unsaturated tris-complexes can be
obtained. This dissociation of a phosphorus ligand as the result of steric
effects has been observed with several transition-metal complexes, mainly
with triphenylphosphine as the ligand, e.g. $Pt(PPh_3)_4$, $RhCl(PPh_3)_3$ and
$RuCl_2(PPh_3)_4$, and as the dissociation is reversible it is encouraged by the
removal of the free ligand from the equilibrium system. One way of doing this
is to use reverse osmosis, a technique in which the solution containing the
ligand and associated metal complexes is forced against a selectively permeable
membrane through which only the solvent and free ligand can diffuse.

The formation of co-ordinatively unsaturated complexes by dissociation
of a ligand, and the shielding by bulky ligands of additional co-ordination
sites on a metal, are two closely related topics both of which are of considerable current interest. The former is important in connection with those
reactions that are catalysed by transition metals, such as hydrogenation and
carbonylation which involve co-ordinately unsaturated complexes (see
chapter 5), while the latter is of interest in current attempts to obtain chemical
models for haemoglobin. In this iron(II) porphyrin one of the axial co-ordination positions is occupied by a globin histidine nitrogen atom, the other by
water, and it is the latter ligand that is displaced by dioxygen in the reversible
formation of oxyhaemoglobin. One of the difficulties encountered during
studies with chemical models of haemoglobin is in the preparation of iron(II)
porphyrins that bear only one nitrogen ligand, for directly one such ligand has
been added the iron changes from high spin to low spin, and its affinity for
another nitrogen ligand is strongly enhanced. This means that in practice one
has the choice of either adding two nitrogen ligands to the iron or none at all!
Another difficulty is to prevent the iron in an iron(II) porphyrin–dioxygen
complex from being irreversibly oxidised to iron(III), a process that involves
a bimolecular reaction with an un-oxygenated iron(II) porphyrin. Both these
difficulties are overcome in haemoglobin by the steric restraints imposed by
the globin residue, for this not only co-ordinates with the iron *via* a histidine
nitrogen but it also sterically restricts access to the other side of the porphyrin
ring and only relatively small ligands such as water, dioxygen, and cyanide and
sulphide anions, can occupy the remaining co-ordination site.

The most successful model so far (Collman, 1977) incorporates a porphyrin

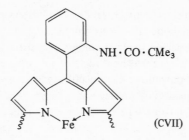 (CVII)

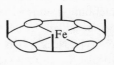 (CVIII)

that bears *o*-pivalamidophenyl groups at each of the four *meso* positions (see partial structure, CVII). The bulkiness of the $Me_3C.CO.NH$ residues not only prevents the benzene rings from being coplanar with the porphyrin ring, but also considerably restricts free rotation about the carbon–carbon bonds connecting these benzene rings to the *meso* carbons. As a result, this porphyrin can exist as four atropisomers which differ from each other in the order in which the $Me_3C.CO.NH$ groups appear on the same side of the porphyrin ring. One of these isomers, of course, has all four groups on the same side of the ring, and this side is therefore sterically hindered to some extent (see sketch CVIII, which also visually explains the use of the colloquial term 'picket-fence porphyrin'). In the presence of nitrogen ligands such as *N*-methylimidazole that preferentially co-ordinate with the iron on the less hindered side of the porphyrin ring, dioxygen can be reversibly added to the remaining co-ordination position to give a stable complex which contains a bent Fe—O—O bond as in oxy-haemoglobin. Steric hindrance by the bulky $Me_3C.CO.NH$ groups inhibits the addition of ligands substantially larger than dioxygen and also considerably retards the oxidation of the iron by the bimolecular route.

So far the discussion has concerned the effect of steric hindrance on the dissociation and addition of monodentate ligands. Now, chelating ligands are bonded to a metal more strongly than the corresponding monodentate ones, and therefore more steric strain is required both to dissociate them from the metal and – what is basically the same thing – to inhibit them from entering into complex formation. As a result, while complexes of monodentate tertiary amines are relatively scarce on account of their intrinsic instability which results from steric factors, many stable complexes of chelating tertiary amines such as N,N-dimethylglycine and 1,2-bis(dimethylamino)ethane are known. The steric strain associated with these chelating ligands becomes apparent, however, when stability constants for complex formation are measured. For example, in aqueous solution both the ligands mentioned above form 1 : 1 chelates with Cu^{2+} and Ni^{2+} which are less stable (by a factor of about 10) than those formed by the parent ligands, glycine and 1,2-diamino-ethane, and this can be attributed in part to steric interactions between the N—Me groups and the co-ordinated water molecules. In fact, the chelating ability of 1,2-diaminoethane is progressively reduced as the four amino hydrogens are systematically replaced by methyl groups. Further evidence of the steric interaction associated with chelated tertiary amines is the very small number of isolable complexes that contain two such ligands bonded to the same metal.

Related to this last point is the observation that steric interaction between two bidentate ligands is usually most pronounced in chelates where, as the result of the stereoelectronic requirements of the metal, the ligand atoms have a square-planar arrangement. This arrangement is possible, of course, not only in a four-co-ordinate complex, but also in a five- or six-co-ordinate complex if the two chelating ligands occupy the equatorial plane. With this arrangement, really severe steric interaction arises when substituents on one or both ligands project towards the other ligand. The copper(II), palladium(II) and platinum(II) chelates of the methyl substituted ligands (CIX–CXII; R = Me), for example, are severely strained because of interactions involving methyl groups (see, for example, partial structure (CXIII) for the ligands (CIX) and (CX)).

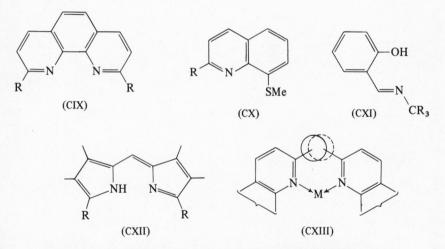

The strain present in copper(II) chelates of the ligands (CIX and CX; R = Me) is largely released when the metal is reduced to copper(I), which unlike the higher oxidation state favours a tetrahedral arrangement of its ligand atoms. This explains why the oxidising power of copper(II) is enhanced more by the presence of 2,9-dimethylphenanthroline (CIX; R = Me) than by the presence of the parent system (CIX; R = H) (cf. p. 31).

One general observation concerning steric strain in metal chelates is that it is generally accommodated by a number of small changes in bond lengths and angles throughout the entire chelate, rather than by considerable distortion in one specific part. In the strained chelates mentioned directly above, one of these changes involves a displacement of the ligand atoms from the desired square-planar arrangement to a distorted tetrahedral one. This results in a hypsochromic shift of the absorption maxima due to the metal d–d transitions, and the colour of the copper(II) chelate of the ligand (CXI; R = Me) for example, is maroon rather than the green exhibited by square-planar copper(II) chelates, e.g. that formed by the unsubstituted ligand (CXI; R = H).

2 Effects and uses of complex formation

2.1. Some general effects of complex formation

In chapter 1 some of the general features of metal complexes were briefly
introduced. The remainder of this book is devoted to the numerous ways in
which metal complexes can be involved in organic chemistry, with particular
emphasis on reactions of synthetic importance. On account of the immensity
of the subject a considerable degree of selectivity has been necessary, but
even so the range of topics considered is a very wide one. It is appropriate,
therefore, first of all to list those properties of metal complexes and those
effects associated with the co-ordination of metals and ligands that cause
metal complexes to be of such importance to the modern organic chemist.
These properties and effects are generally the result of one or more stereochemical, kinetic and thermodynamic factors, and they may be summarised
as follows:

1. The physical properties of metal complexes are markedly different
from those of the unco-ordinated ligands. This fact often permits ligands to
be isolated or purified or both, *via* metal complexes (§2.3).

2. By co-ordinating with one or more ligands that are components of an
equilibrium system, a metal can alter the equilibrium position of that system.
The presence of a metal will often allow a thermodynamically unfavourable
reaction to proceed to completion (§2.7). The same effect also accounts for
the increased acidity of certain co-ordinated ligands (§2.8).

3. The arrangement of the ligand atoms about the metal in a complex is
usually highly stereospecific. Accordingly, the co-ordination of a chelating
ligand reduces the range of conformations that the ligand can adopt (§2.2).
This can affect the reactivity of the ligand, particularly in those reactions that
require the ligand to adopt specific conformations, e.g. in the synthesis of
macrocycles.

4. If a metal complex is not chiral as the result of the stereochemical
arrangement of the ligands about the metal, it can be made so by introducing
elements of chirality into one or more of the ligands. Chiral metal complexes

can be used for resolving racemic ligands (§2.4). Any synthesis that involves metal complexes can be made to proceed in an asymmetric manner by incorporating chiral ligands into the complex (see, for example, asymmetric catalytic hydrogenations, §5.5).

5. Co-ordination of an unsaturated ligand by means of the π-electrons associated with the unsaturation often results in a complex whose chemical properties are considerably different from those of the free ligand. For example, the complex may be far more stable with respect to hydrolysis, oxidation, and thermal degradation; many complexes of highly transient species such as carbenes and cyclobutadiene can be isolated and stored at room temperature.

6. Co-ordination of a ligand by means of a lone pair of electrons on the ligand atom reduces the basicity and nucleophilicity of that atom. This may result in protonation or attack by an electrophile taking place at an alternative site in the ligand (§2.5).

7. If in an unco-ordinated ligand a lone pair of electrons is involved in resonance stabilisation, then co-ordination of the ligand by means of that lone pair will inhibit the resonance stabilisation. This in turn will increase the reactivity of the ligand in reactions where resonance stabilisation is lost in the rate-determining step, for example, in the hydrolysis of carboxylic acid derivatives (§3.11). Co-ordination will also result in a reduction in the basicity and nucleophilicity of those atoms in the ligand on which negative charge would normally accumulate as the result of the resonance stabilisation.

8. Co-ordination of a ligand with a metal cation can result in polarisation of bonds within the ligand. In extreme cases heterolytic fission of bonds may occur. The polarisation resulting from co-ordination can enhance the rate of attack on the ligand by nucleophiles. This enhancement is particularly high if the negative charge which accumulates in the ligand as the result of nucleophilic attack is situated on one of the co-ordinated ligand atoms as, for example, in the hydrolysis of α-amino esters (§3.12) and in nucleophilic attack on co-ordinated alkenes (§4.4).

9. Back-bonding of electrons from a metal can increase the basicity and nucleophilicity of those ligands that accept the electrons. See for example, the increased basicity of co-ordinated pyrazine (p. 29).

10. Co-ordination allows functional groups in adjacent ligands to react intramolecularly. The reaction therefore generally proceeds faster than the corresponding intermolecular reaction that would take place in the absence of the metal, because there is a reduction in the entropy loss associated with formation of the transition state.

11. In some reactions between adjacent ligands, one or more metal–ligand bonds are involved. Examples include not only those reactions where the metal serves as a carrier for a nucleophilic species, e.g. alkyl, hydride and hydroxyl anions, but reactions where an unsaturated ligand is inserted into an

adjacent metal–ligand bond, e.g. the insertion of alkenes (§5.4) and carbon monoxide (§5.7) into metal–hydrogen and metal–carbon bonds.

12. Electron transfer between a metal and its ligands can result in reduction of the metal and oxidation of one or more ligands, or *vice versa*. Metal complexes are often involved in reactions where a metal cation oxidises an organic compound.

13. With some metal complexes the metal can undergo a reductive elimination, i.e. the metal undergoes a formal two-electron reduction, and two σ-bonded ligands become linked together (§5.1). This type of reaction provides a very convenient route for linking together two residues that can be bonded individually to a metal (see synthetic applications, §5.2).

14. Some metal complexes – notably those in which the metal is in a low oxidation state – react with other compounds by the metal inserting itself into a single bond in one of the reactants. This results in the creation of two new ligands on the metal which are available for further reaction. A good example of this is in homogeneous hydrogenation (§5.5) where two hydride ligands are created by a metal inserting itself into the H—H bond of a hydrogen molecule.

A very high proportion of the uses and reactions of metal complexes described in the remainder of this book can be analysed in terms of one or more of the effects and properties listed above.

2.2. Stereochemical effects of complex formation

One stereochemical result of the co-ordination of a monodentate ligand with a metal is the change in the relative stabilities of the various conformations which the ligand can assume, because of the increase in the effective size of the co-ordinating group. In cyclohexane systems, for example, the stabilities of conformations that have an equatorial hydroxyl, alkoxy, or amino group, are increased when that group participates in complex formation. This effect of complex formation on the equilibria between the various conformations of an organic ligand should always be borne in mind when n.m.r. shift reagents are used in stereochemical studies, for co-ordination of the ligand with the shift reagent may affect not only the n.m.r. spectrum of the former but also its stereochemistry (see discussion by Hofer, 1976).

Although this steric effect is also present in complexes of di-, tri- and multidentate ligands, it is usually masked or even inhibited by the fact that the range of conformations which the ligand can adopt is severely restricted if all the co-ordination centres of the ligand are involved in co-ordination. The most stable conformation of ethylenediamine is the antiperiplanar one (I), and it is to be expected that the relative stability of this conformation will be increased if the two amino groups are individually co-ordinated to two metal ions. Nevertheless, this conformation cannot be involved in chelates in which both amino groups are co-ordinated with the same metal ion.

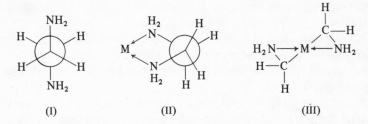

(I)   (II)   (III)

From an inspection of torsional and bending energy terms Corey and Bailar (1959) initially suggested that chelates of this diamine exis ted as two skew enantiomeric conformers, e.g. (II), in which the two carbon atoms were equidistant from, but on opposite sides of, the plane of co-ordination (see III), and that the dihedral angle was about 49°. From X-ray data it is now evident that in fact the five-membered ring does exist in this type of unsymmetrical conformation, and that the dihedral angle is in the range 45–55°. Chelates of related ligands of the 1,2-disubstituted ethane type, e.g. ethanedithiol, must also be similarly restricted in the number of conformations they may adopt. Analogously, in metal complexes of conjugated dienes, e.g. (IV), the diene system is held in the *s*-cisoid conformation, which in the free ligand is about 9.5 kJ mol^{-1} less stable than the alternative *s*-transoid conformation.

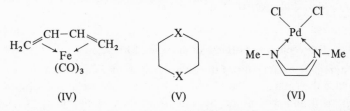

(IV)   (V)   (VI)

Infrared studies of complexes of the cyclic systems (V; X = O, S and NH) show that the ligand is in the chair form, and the two heteroatoms are co-ordinated to different metal ions. With palladium(II) chloride, however, the heterocycle (V; X = NMe) acts as a bidentate ligand and gives a 1 : 1 chelate in which the heterocyclic ring is held in the boat conformation (see VI). This conformation is found in many chelates of the related tetra-amine (V; $X = N.(CH_2)_3.NH_2$) which can function as a tetradentate chelating ligand.

Conformations of monosaccharides with an *ax–eq–ax* sequence of three hydroxyl groups around the ring are stabilised by the presence of alkaline-earth metal ions, because this arrangement of the three hydroxyl groups allows the formation of stable chelates. Thus in aqueous calcium chloride, methyl β-D-ribopyranoside exists largely as (VII). Similarly, with many metal ions, *cis,cis*-1,3,5-triaminocyclohexane forms stable chelates (VIII) in which all three amino groups are axial as shown. In both of these systems the co-ordinated ligand is in a conformation which in the absence of the metal would be an unfavourable one.

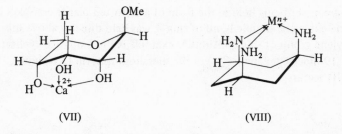

(VII)   (VIII)

The ability of a metal ion to induce multidentate ligands to adopt energetically unfavourable conformations has been used to a small extent in synthetic work, particularly in the synthesis of macrocyclic chelating agents. Usually the formation of a macrocycle by the reaction between two functional groups at the ends of a linear, acyclic precursor proceeds at very slow rate because of an unfavourable entropy factor which results from the precursor having to adopt an almost cyclic conformation for the two functional groups to come close enough to react. If the precursor can chelate with a metal ion, however, the former may be held in a conformation in which the two functional groups are in close proximity, and macrocycle formation will then readily occur around the metal ion. Subsequent removal of the metal ion from the resultant chelate affords the free macrocyclic ligand, but it should be noted that with some systems the chelate may be so unreactive that it is only possible to remove the metal ion by use of vigorous conditions which also cause degradation of the ligand.

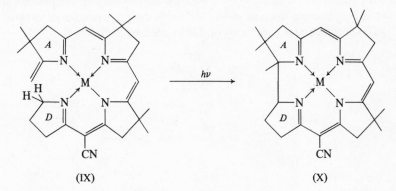

(IX)   (X)

The chelating properties of porphyrins and corrins have allowed several of these macrocyclic systems to be synthesised by the technique described above. In a very elegant corrin synthesis discussed by Eschenmoser (1970), photochemical irradiation of the complex (IX; M = Pd^+, Pt^+, or MgCl) caused a 1,16 antarafacial transfer of hydrogen and cyclisation to the corresponding corrin (X); the removal of the magnesium from (X; M = MgCl) afforded the metal-free system. The success of this cyclisation was largely due to the tetrapyrrole precursor being held in the form of a distorted planar complex in which the exocyclic double bond of ring *A* was held directly above the *N*-methylene group of ring *D*. Another example is the oxidative cyclisation of the 1,19-dideoxybilene-*b* (XI) to the metalloporphyrin (XII) with copper(II) acetate.

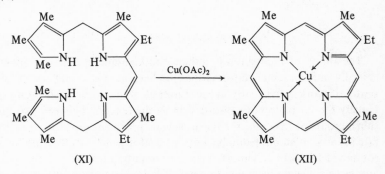

(XI)   (XII)

The holding of a multidentate ligand in a specific conformation occasionally accounts for the fact that in the presence of metal ions a reaction will proceed along a pathway which in the absence of the metal is only a minor contributor. A synthetic example of this is the preparation of 2,5-dihydroxy-benzoquinone by an intramolecular Claisen condensation with methyl 2,4,5-trioxohexanoate. With sodium alkoxides, tertiary amines, or quaternary ammonium salts as the basic reagent, the quinone is only a very minor product. With magnesium methoxide, however, excellent yields of the quinone are obtained, because the magnesium ion holds the precursor (as an enolate – see XIII) in the correct conformation for the desired cyclisation.

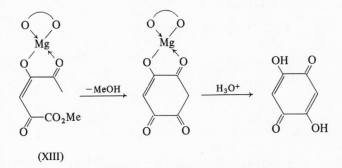

(XIII)

In connection with this reaction it should be noted that when treated with magnesium methoxide a number of poly-β-carbonyl compounds give products by base-catalysed cyclisations which are completely different from those that occur by use of sodium methoxide (e.g. transformations of XIV), and which can be rationalised in terms of magnesium chelates. As these magnesium-induced cyclisations are of the type believed to take place in the biosynthesis of phenolic compounds from acyclic poly-β-carbonyl precursors, it has been suggested that metal chelates may be involved in the biosynthetic processes.

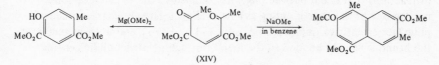

(XIV)

As a result of the ligand being held in a particular conformation the reactions of a chelated ligand are often highly stereoselective. The addition of aromatic Grignard reagents to the complex formed from tin(IV) chloride and 2-amino-2-phenylacetophenone, for example, is 98 per cent stereo-selective and gives alcohols having the stereochemistry shown in (XVI), because in the complex the aminoketone is rigidly held in the form of the five-membered ring chelate (XV), one side of which is far more accessible to the Grignard reagent than the other.

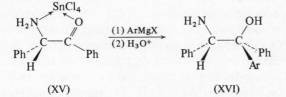

(XV)   (XVI)

The ligands discussed so far have been comparatively simple in structure, and it must be pointed out that with multidentate ligands that have a very large number of co-ordinating groups, e.g. peptides, not all metal ions will cause the ligand to adopt the same conformation. Because of the variations in the stereochemical and electronic requirements of individual metal ions, completely different sets of co-ordinating groups may co-ordinate with different metal ions. This fact might explain the specificity of the requirements of certain enzymes for metal ions, for if the enzyme has to adopt a highly specific conformation in order to react with the substrate, it seems feasible that this might result only from the enzyme co-ordinating with a metal whose dimensions and stereochemistry exactly coincide with the co-ordination characteristics of the required conformation. This explanation, together with other possible roles of metal ions in enzyme systems, has been discussed by Gillard (1967).

In the synthesis of macrocycles by the technique outlined earlier, the cyclisation step is an intramolecular reaction between two reactive centres in the same ligand. Many metal-promoted transformations involve reactions between reactive centres in different ligands. In these transformations the two ligands are invariably adjacent to each other in the co-ordination sphere of the metal, and usually the reaction between them proceeds through a transition state with a five- or six-membered ring. In some cases the reaction products remain co-ordinated with the metal, while in others they are displaced, thus making the metal available for further reaction. The majority of reactions that occur between ligands attached to the same metal ion involve one of the ligands acting as a neutral or negatively charged nucleophile (for specific examples see pp. 148, 171 and 174), and in these reactions the ligand is usually bonded to the metal through its nucleophilic centre.

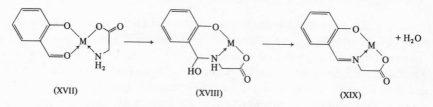

One interesting feature of these interligand reactions is that sometimes they take place less rapidly when the ligands are co-ordinated to a transition metal than they do when a main group metal is involved. Kinetic studies of the metal-promoted formation of the chelates (XIX) from salicylaldehyde, glycine and different metal ions have shown that a rapid equilibrium to give the mixed chelate (XVII) is followed by rate-determining attack of the $-NH_2$ group on the co-ordinated aldehyde, and dehydration of the resultant amino alcohol (XVIII). The rate-determining interligand reaction is very much slower when M is Cu, Ni and Co than when Pb, Cd, Mg and Zn are used. A likely explanation for this is that as transition-metal ions have unfilled d-orbitals and the associated ligands are strongly bound to the metal at co-ordination sites having a well-defined geometry, the reactive centres in the ligands of the mixed complex (XVII) are widely separated when M is a transition metal. The interligand reaction cannot therefore proceed until sufficient energy is supplied for the ligands to assume a more favourable but less stable configuration around the metal. This effect is not nearly as pronounced with metal ions which have filled d-orbitals, and hence the reaction proceeds more readily.

In interligand reactions the ligands concerned often become bonded to each other, e.g. (XVII → XIX). In extreme cases a consecutive series of reactions leads ultimately to all of the ligands around the metal becoming bonded to each other, and the final product of the reaction is a macrocylic ligand with the metal ion held in the middle of the ring. With dry acetone, bis(ethylenediamine)nickel(II) perchlorate affords the chelate (XX) by a series of interligand reactions which involve imine formation and base-catalysed additions of $-CH_3$ groups to C=N bonds.

It is not surprising that when a macrocylic chelating agent is synthesised by a reaction that involves multiple condensations, the yield of macrocycle is often increased at the expense of linear polycondensation products when the reaction is carried out in the presence of metal ions. Thus the addition of

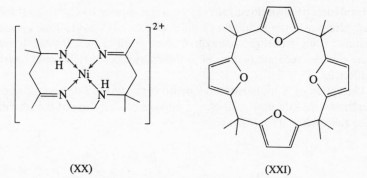

(XX)   (XXI)

Li^+, Ca^{2+}, Zn^{2+} or Mg^{2+} to the reaction mixture increases from 18-20 per cent to 40-45 per cent the yield of the macrocycle (XXI) obtained from furan and acetone in acidified ethanol.

The observation that metal ions can act as templates and promote cyclisations that lead to macrocycles has led to these cyclisations being termed 'template reactions'. This term is now also frequently applied to any reaction between two or more ligands attached to the same metal ion. Busch (1967) has reviewed template reactions that lead to macrocyclic ligands, and much additional information on this topic can also be found in the reviews on macrocyclic ligands by Christensen and Izatt (1978) and Lindoy (1975).

2.3.   Isolation and purification of organic ligands

When an organic ligand co-ordinates with a metal, the resultant complex always has some physical properties that are completely different from those of the two components. The Mond process for the purification of metallic nickel *via* the highly volatile tetracarbonylnickel(0) is an example of the commercial exploitation of this fact. Many organic ligands can similarly be isolated and/or purified by converting them into a metal complex, which if necessary can be purified by a suitable method, e.g. crystallisation, and from which the ligand can subsequently be regenerated – usually by treating the complex with another ligand that has a higher affinity for the metal than the ligand that is to be recovered. This technique not only permits the separation of organic ligands from compounds that have no complexing ability, but also enables two or more ligands to be separated by virtue of the difference between their affinities for the same metal ion. The following examples illustrate this technique.

At −15 °C copper(I) chloride forms a complex with styrene from which the styrene can be recovered by simply warming the complex to about 35 °C. As ethylbenzene does not complex with copper(I), styrene contaminated with ethylbenzene is easily purified by treating the crude material with copper(I) chloride and subsequently warming the resultant complex.

Oxidation of phenylhydrazine by aqueous copper(II) chloride affords phenyldiazene in the form of a complex $(PhN=NH)Cu_4Cl_4$ which is readily separated from unchanged phenylhydrazine by filtration. The highly reactive diazene can be recovered from the complex by ligand displacement with acetonitrile.

The action of *N*-bromosuccinimide on cyclo-octa-1,3-diene followed by reduction with lithium aluminium hydride gives a mixture of the starting diene and cyclo-octa-1,4-diene.

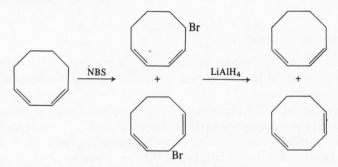

The former alkene is inert to silver nitrate, but the latter forms a water-soluble complex and is therefore conveniently separated from the conjugated isomer by extraction into aqueous silver nitrate, followed by the addition of ammonia.

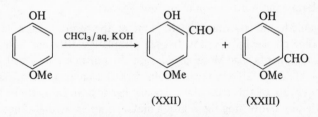

An example of the separation of a chelating agent from a complexing agent is provided by the Reimer–Tiemann reaction with 4-methoxyphenol which gives a mixture of the two isomers (XXII) and (XXIII). Only the former gives a stable copper(II) chelate (cf. p. 5) and this is precipitated quantitatively when the mixture is treated with aqueous copper(II) acetate. Decomposition of this chelate by hydrochloric acid affords pure 5-methoxysalicylaldehyde (XXII).

The two stereoisomers (XXIV) and (XXV) formed in approximately equal amounts when the trioxime of cyclohexan-1,3,5-trione is reduced by Birch's method can also be conveniently separated by the action of metal ions. The *cis,cis* isomer (XXV) can function as a tridentate ligand (see (VIII), p. 77), and hence its chelates are far more stable than those of the *cis,trans* isomer (XXIV), whose stereochemistry only allows it to function as a bidentate

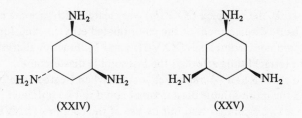

(XXIV)   (XXV)

ligand. Addition of the mixture of stereoisomers to an aqueous solution of a limited amount of nickel(II) nitrate therefore preferentially gives the cation (VIII; M = Ni^{2+}) as the dinitrate salt. This separates as a crystalline solid from which the triamine can be obtained by treatment with hydrochloric acid.

2.4. Resolution of racemic mixtures

In addition to their use for separating ligands, metal complexes have occasionally been used for effecting the separation of the enantiomers of the same ligand, i.e. the resolution of a racemic mixture. The technique usually employed is to combine the racemic ligand (±)-L with a chiral metal complex, say (+)-M, to produce a mixture of two diastereomeric products (+)-L/(+)-M and (−)-L/(+)-M which are separated by fractional crystallisation. The two products are then individually treated with a reagent which will displace the ligand (L) without causing racemisation. The metal complex used may be chiral either because of the stereochemical arrangement of the ligands around the metal, or because the complex contains a chiral ligand. These two situations are illustrated by resolution of (±)-2,3-dimethylsuccinic acid *via* salt formation with the dextrorotatory tris(ethylenediamine)cobalt(III) cation (XXVI), and the resolution of *trans*-cyclo-octene by use of the enantiomeric complexes (XXVII; *a* and *b*).

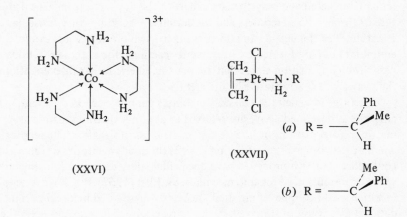

(XXVI)

(XXVII)

(*a*) R =

(*b*) R =

In the latter work, the complex (XXVII*a*) was treated with racemic *trans*-cyclo-octene to cause displacement of the co-ordinated ethylene and formation of the two diastereomeric complexes (XXVIII; *a* and *b*). Fractional crystallisation from carbon tetrachloride afforded the less soluble diastereomer (XXVIII*a*) which gave (R)- or (-)-*trans*-cyclo-octene when treated with potassium cyanide. As the more soluble diastereomer could not be obtained pure, the racemic alkene was again resolved but by use of the complex (XXVII*b*). The same complexes (XXVII; *a* and *b*) have also been used for resolving other ligands that can displace the co-ordinated ethylene and hence form a mixture of diastereomers, e.g. *cis,trans*-cyclo-octa-1,5-diene, ethyl *p*-tolylsulphoxide and cyclonona-1,2-diene.

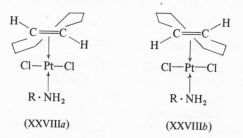

(XXVIII*a*)   (XXVIII*b*)

A related method which uses metal complexes for resolving racemic ligands is based on the difference between the thermodynamic stabilities of the two diastereomers produced when a chiral complex reacts with a racemic ligand as described above. If the two diastereomers can equilibrate (either directly or through the starting chiral complex), their relative proportions will be determined by the free energy difference between them. If this difference is very large and two equivalents of the racemic ligand are used in the reaction, one of the diastereomeric products will be formed exclusively from the appropriate enantiomeric ligand, while the other enantiomeric ligand will remain unreacted and may then be isolated. The deviation from a 1 : 1 mixture of the two diastereomers and the selective reaction of one particular enantiomer of the ligand can also result if the two diastereomers cannot equilibrate but are formed by irreversible reactions, for as these reactions pass through transition states that are diastereomeric rather than enantiomeric their rates will nearly always be different.

Reactions in which the two diastereomers are produced in different amounts as described above are referred to as either thermodynamic asymmetric transformations or kinetic asymmetric transformations, depending on whether the composition of the product is thermodynamically or kinetically controlled. The two processes have been adequately discussed in connection with the resolution of organic racemates by Eliel (1962), but some examples in which metal complexes are used are briefly mentioned below. For a more detailed discussion of stereoselectivity in reactions of metal complexes the reader is referred to reviews by Dunlop and Gillard (1966), and Gillard (1967).

With some reversible systems, for example the formation of Cu[(L)-alanine]$_2$ and Cu[(L)-alanine] [(D)-alanine] from [Cu(L-alanine)]$^+$ and DL-alanine, no thermodynamic stereoselectivity can be detected, but others show partial stereoselectivity. In the formation of the cobalt(II) and nickel(II) complexes of histidine, the racemic species M[(D)-histidine] [(L)-histidine] are more stable than the optically pure species M[(D)-histidine]$_2$ and M[(L)-histidine]$_2$, with differences in $\Delta G°$ of about 1.3 and 2.1 kJ mol^{-1} respectively. Thermodynamic stereoselectivity is also exhibited in the displacement of the co-ordinated ethylene from the platinum(II) complexes (XXVII, *a* and *b*) by *trans*-cyclo-octene. When the 1-phenylethylamine has the S-configuration as in (XXVII*b*), the diastereomeric product (XXVIII*b*) with the (+)- or S-form of the cyclo-octene is the more stable, and so if the complex (XXVII*b*) is treated with two equivalents of racemic *trans*-cyclo-octene the excess of the alkene is enriched in the (–)-enantiomer and has an optical purity of about 30 per cent. Partial resolution is also observed in the reaction of the same complex with racemic alkenes of the type RR'CH . CH=CH$_2$.

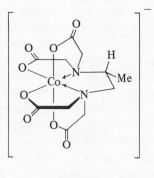

(XXIX)

An excellent example of the resolution of an organic racemate by means of a kinetic asymmetric transformation is the displacement of (+)-propylene-diaminetetra-acetate anion from the anion (XXIX) by propylenediamine (pd). This reaction has a very high degree of stereoselectivity, and if a 100 per cent excess of the racemic diamine is used only the (+)-enantiomer is incorporated into the product:

$$\{Co[(+)\text{-PDTA}]\}^- + 6\,(\pm)\text{-pd} \rightarrow \{Co[(+)\text{-pd}]_3\}^{3+} + 3(-)\text{-pd} + \text{PDTA}^{4-}$$

Lysine may be resolved similarly by use of the analogous laevorotatory cobalt(II) complex of EDTA, which selectively forms a salt at pH = 4 with the laevorotatory enantiomer of the amino acid:

$$(-)\text{-}[Co(EDTA)]^- + (\pm)\text{-Lysine} \rightarrow \{(-)\text{-Lysine}/(-)\text{-}[Co(EDTA)]^-\} + (+)\text{-Lysine}$$

In addition to their use as resolving agents, chiral metal complexes also appear to be very promising reagents for achieving asymmetric syntheses. A number of these syntheses are described elsewhere in this book, e.g. pp. 152 and 253.

2.5. Metals as protecting groups

The changes in electron densities within a ligand that occur when the ligand co-ordinates with a metal may result in the ligand being activated towards attack by one particular type of reagent, but deactivated towards attack by a different type of reagent. Thus co-ordination may facilitate reactions between the ligand and nucleophiles, but retard reactions with electrophiles. Although activation of the ligand is undoubtedly the more exploited of the two effects, deactivation has often been utilised in organic synthesis, mainly in the protection of nucleophilic functional groups. When such groups are co-ordinated with a metal ion by acting as donor ligands, their ability to function as nucleophilic centres for electrophilic reagents is substantially reduced.

The α-carboxylate group of the dianion of glutamic acid (XXX; $n = 2$, $X = CO_2H$), is less nucleophilic than the γ-carboxylate group when only the former is co-ordinated to a metal ion as in the copper(II) chelate (XXXI; $n = 2$, $X = CO_2^-$). Consequently when this chelate is treated with benzyl chloride, ester formation occurs selectively at the γ-group, and subsequent removal of the copper by precipitation as the sulphide constitutes the last stage of a convenient synthesis of γ-benzyl glutamate. The nucleophilicity of an amino group is similarly reduced by co-ordination, and reactions between the copper(II) chelates of lysine (XXX; $n = 4$, $X = NH_2$) and ornithine (XXX; $n = 3$, $X = NH_2$) and electrophilic reagents such as acetic anhydride and benzyloxycarbonyl chloride involve almost exclusive attack on the unco-ordinated ω-amino group. For the same reason, citrulline (XXX; $n = 3$, X $= NH.CO.NH_2$) may be conveniently obtained in good yield from the copper chelate (XXXI; $n = 3$, $X = NH_2$) and urea, and tyrosine may be selectively *O*-benzylated by treating the copper chelate of this amino acid with benzyl chloride.

$$X \cdot (CH_2)_n \cdot CH(NH_2)CO_2H$$

(XXX)

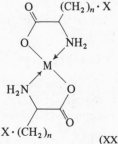

(XXXI)

The reduced nucleophilic character of a co-ordinated amino group was utilised by Eschenmoser and his workers in their elegant synthesis of the corrin ring system (Eschenmoser, 1970). Reaction of the silver salt (XXXII) with triethyloxonium tetrafluoroborate afforded the desired *O*-alkylated product (XXXIII), in contrast to the reaction with the parent ligand which resulted in indiscriminate ethylation at oxygen and nitrogen.

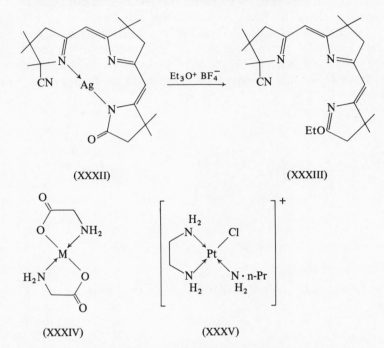

(XXXII)   (XXXIII)

(XXXIV)   (XXXV)

Although the above-mentioned reactions demonstrate that co-ordination effectively reduces the nucleophilic character of an amino group, there is some evidence that suggests that a co-ordinated primary amino group can still function as a nucleophilic centre. The reaction between acetaldehyde and the glycine chelate (XXXIV; M = Cu) to give the analogous threonine chelate appears to involve the initial formation of a Schiff base (see p. 120), and both amino groups of the copper chelate (XXXIV; M = Cu) may be monocyanoethylated by the action of acrylonitrile; in the presence of an excess of the nitrile the corresponding palladium(II) chelate (XXXIV; M = Pd) gives the N,N,N',N'-tetrasubstituted derivative. Similarly, treatment of an aqueous solution of the copper chelate (XXXI; $n = 4$, X = NH_2) with ethylene oxide results in the introduction of a 2-hydroxyethyl residue into the α-amino group as well as into the γ-amino group. The amine group of the co-ordinated propylamine in the cation (XXXV) may be nitrosated to give the expected *N*-nitroso derivative.

$$R-\underset{H_2}{N}: \longrightarrow M^{n+} \rightleftharpoons R-\underset{H}{\overset{..}{N}}-M^{(n-1)+} + H^+ \qquad [2.1]$$

These results can be explained by assuming that there is appreciable dissociation of the chelates concerned, and that the resultant unco-ordinated amino group then functions as a nucleophile in the usual manner. An alternative and, in most cases, more plausible explanation is that on account of its increased acidity (see §2.8) the co-ordinated amino group is deprotonated to a small extent (equation 2.1), and the co-ordinated conjugate base is the real nucleophilic species of the reaction. This explanation is supported by the ability of certain ammine complexes to react with electrophilic reagents under basic conditions that promote deprotonation but not displacement of the ammonia ligands:

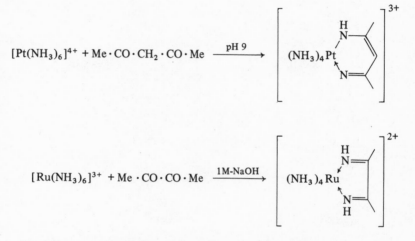

$$[Pt(NH_3)_6]^{4+} + Me \cdot CO \cdot CH_2 \cdot CO \cdot Me \xrightarrow{pH\ 9}$$

$$[Ru(NH_3)_6]^{3+} + Me \cdot CO \cdot CO \cdot Me \xrightarrow{1M\text{-}NaOH}$$

In this context it should be noted that even though the negative charge of the conjugate base is delocalised by the metal ion, complexes in which a deprotonated amine is one of the ligands are extremely effective nucleophiles (e.g. equation 2.2).

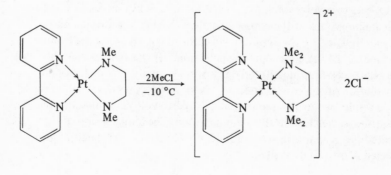

$$[2.2]$$

In metal–alkene complexes the overlap of the filled π-orbital with an empty orbital on the metal prevents the alkene from entering into reactions that normally require the participation of the π-electrons. Haloketones of the type (XXXVI), for example, readily undergo nucleophilic substitutions by the mechanism shown in equation 2.3,

$$R{-}\underset{\underset{O}{\|}}{C}{-}CH{=}C(X)R' \xrightarrow{Nu^-} R{-}\underset{\underset{O^-}{|}}{C}{=}CH{-}\underset{\underset{X}{|}}{\overset{\overset{Nu}{|}}{C}}{-}R' \longrightarrow R{-}\underset{\underset{O}{\|}}{C}{-}CH{=}C(Nu)R' + X^- \qquad [2.3]$$

(XXXVI)

but this mechanism is prevented if the double bond is co-ordinated to a metal, and consequently the halogen in the complex (XXXVII) is relatively inert to nucleophilic displacement compared with that of the uncomplexed ligand.

$$Ph\cdot CO\cdot \underset{\underset{Fe(CO)_4}{\downarrow}}{CH{=}CHCl}$$

(XXXVII)

Similarly, co-ordination of a double bond with an $[Fe(\eta\text{-}C_5H_5)(CO)_2]^+$ residue deactivates the double bond towards electrophilic additions, thus allowing molecular bromine, for example, selectively to substitute the aromatic ring of eugenol (XXXVIII) rather than adding to the reactive double bond in the side-chain. The residue can be introduced by displacement of the methylpropene from $[Fe(\eta\text{-}C_5H_5)(CO)_2(Me_2C{:}CH_2)]^+$, and removed by the action of sodium iodide in acetone.

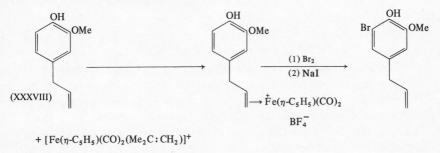

Generally, alkenes co-ordinated to Ag^+ cannot be catalytically hydrogenated, and consequently a functional group such as an oxirane ring can often be selectively reduced in the presence of a carbon–carbon double bond if the latter is protected as a silver complex; in practice the complex is conveniently formed *in situ* by the addition of silver nitrate to the hydrogenation medium. It should be noted however, that some co-ordinated alkenes can be successfully reduced (see equation 2.4). This is probably because the metal involved can react with hydrogen – either directly or after reduction to a lower oxidation state – and form the hydride species essential for the hydrogenation of the alkene (see §5.5).

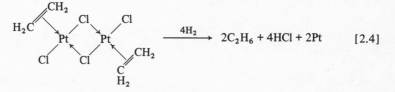

$$\xrightarrow{4H_2} 2C_2H_6 + 4HCl + 2Pt \qquad [2.4]$$

   Conjugated dienes may be protected against catalytic hydrogenation by co-ordination to an iron tricarbonyl residue. This co-ordination, in fact, protects the diene system against a variety of reagents, and the isolated double bond in the myrcene complex (XXXIX), for example, can successfully be reduced, hydroboronated, and hydrated with H_3O^+ without modification of the co-ordinated diene. Dienes protected in this manner do not exhibit Diels–Alder reactions, except under forcing conditions which result in dissociation of the complex or displacement of the diene from the metal by the dienophile. The iron tricarbonyl residue can be removed from the diene by the action of inorganic oxidising agents or tertiary amine oxides as described on p. 92.

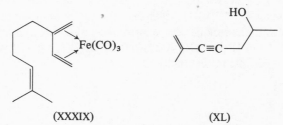

(XXXIX)   (XL)

   When treated with dicobalt octacarbonyl, alkynes form complexes in which the triple bond is co-ordinated with a dicobalt hexacarbonyl residue. This residue is similar to the iron tricarbonyl residue in that not only can it be removed by oxidising agents such as Ce^{4+} and Fe^{3+} but it is stable to a variety of reagents. If the triple bond of the unsaturated alcohol (XL) is protected in this manner, for example, the double bond can be reduced with di-imide, and also hydrated to give the expected primary alcohol by treatment with diborane followed by mild oxidation. The same method of protecting the triple bond in diarylacetylenes, $ArC{\equiv}CAr'$, allows the aryl residues to be acylated by Friedel–Crafts procedures.

   The final examples of co-ordination being used to protect functional groups illustrate the principle that certain groups may be protected against nucleophilic attack by forming a metal complex of the conjugate base.

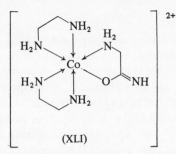

(XLI)

For example, the amide group of glycine amide is completely inert to nucleophilic attack by hydroxyl anions when the conjugate base is chelated with Co^{3+} in the bis(ethylenediamine) cation (XLI).

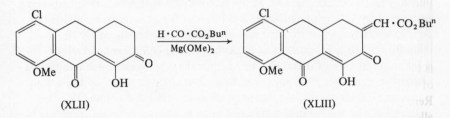

(XLII)   (XLIII)

The lithium or sodium enolate of a ketone is stable to metal hydride reduction, and the chelated anion of a β-diketone is similarly resistant to attack by a wide range of nucleophiles. The latter fact was utilised during the synthesis of ($\pm$)-6-demethyl-6-deoxytetracycline (Woodward, 1968) when it became necessary to carry out the transformation (XLII) → (XLIII) under basic conditions that would not also cause cleavage of the β-dicarbonyl system. This was conveniently achieved by use of magnesium methoxide as the condensing agent, for this actually protected the diketone system by converting it into a stable magnesium chelate.

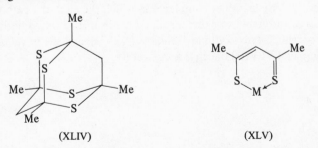

(XLIV)   (XLV)

Dithioacetylacetone ($Me.CS.CH_2.CS.Me$) appears to be a highly reactive compound, for attempts to prepare it have usually resulted in the formation of a dimer (XLIV) that has an adamantane-like structure. In contrast, the anion of the monomer forms stable chelates of type (XLV) with a wide range of di- and trivalent metal cations.

2.6. Stabilisation of reactive species

The change in reactivity which can arise when an organic species is co-ordinated to a metal explains the existence of many stable complexes of species which in the uncomplexed state are very short-lived, on account of their high reactivities. Because of the ease with which the carbon monoxide groups are displaced from pentacarbonyliron(0) by other ligands, some of these unstable species have been generated in the presence of the penta-carbonyl, and then subsequently isolated as iron carbonyl complexes. Ennea-carbonyldi-iron, $Fe_2(CO)_9$, is also frequently used for this purpose, because above 50 °C this compound dissociates into $Fe(CO)_5$ and $Fe(CO)_4$. The last compound is co-ordinatively unsaturated and very rapidly adds an additional ligand; displacement of one or more carbon monoxide groups occurs if the ligand is multidentate. Because of the relative ease with which iron carbonyl complexes may be prepared – and also because they are fairly amenable to purification by column chromatography and crystallisation, singly or combined – a high proportion of the known metal complexes of various reactive species are of this type. An additional advantage of these complexes is that the iron carbonyl residue can be conveniently removed by the action of t-amine oxides or oxidising agents such as iron(III) and cerium(IV) ions. Removal of the residue in the presence of an additional reactant therefore allows a study of the reactions of the uncomplexed reactive species. This is evidently a very convenient technique for examining the chemistry of highly reactive and elusive species.

An example of a ligand which in the uncomplexed state rapidly rearranges to a more stable compound is the co-ordinated vinyl alcohol of the platinum complex (XLVI; R = H), which is obtained as a stable yellow solid, m.p. 127–131 °C, by very mild hydrolysis of the corresponding trimethylsilyl ether (XLVI; R = $SiMe_3$). The co-ordinated vinyl alcohol is displaced from the complex by pyridine, and may then be detected as the more stable tautomer, acetaldehyde. Other complexes in which the enol of a carbonyl compound is co-ordinated to a metal include the iron tricarbonyl derivatives of 2-hydroxy- and 1-hydroxy-buta-1,3-diene (XLVII), and the novel plati-num(II) complex (XLVIII) of ethyl acetoacetate.

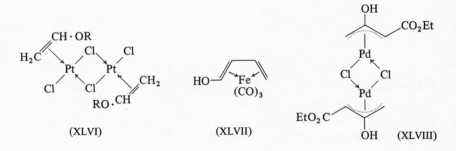

(XLVI)   (XLVII)   (XLVIII)

To complement these co-ordinated enols, the ketone tautomer of phenol has been obtained as the stable iron tricarbonyl complex (XLIX). The carbonyl function ($\nu_{max} = 1663$ cm^{-1}) of the co-ordinated dienone system exhibits the usual reactions with nucleophilic reagents, and may be reduced by sodium borohydride or converted into the expected β-hydroxy-ester by a Reformatskii reaction with methyl bromoacetate without affecting the iron tricarbonyl group. However, this group can be displaced from the complex (XLIX) by the action of iron(III) chloride and the liberated carbocyclic ligand is then observed as phenol.

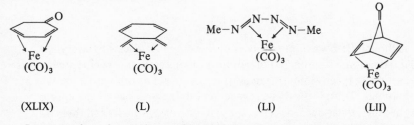

(XLIX)   (L)   (LI)   (LII)

Iron tricarbonyl complexes of ligands that in the uncomplexed state are either completely unknown or are extremely unstable include (L), (LI), (LII), and the complex of the anti-aromatic species, cycloheptatrienyl anion.

Metal complexes of cyclobutadiene and its derivatives (Maitlis, 1966) have attracted much attention since Longuet-Higgins and Orgel suggested in 1956 that this unstable ring system could be stabilised by co-ordination to a transition metal. A number of these complexes have since been prepared by methods which include the cyclodimerisation of alkynes by metal complexes, ring closure of certain 1,4-disubstituted buta-1,3-dienes in the presence of metal salts, and the dehalogenation of dihalocyclobutenes or tetrahalocyclobutanes in the presence of metal carbonyls.

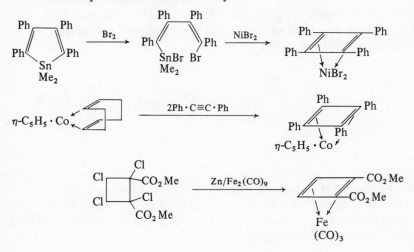

The reaction between *cis*-3,4-dichlorocyclobut-1-ene and enneacarbonyldi-iron in pentane affords the iron tricarbonyl complex (LIII) of the unsubstituted cyclobutadiene system, and the same complex may also be obtained by photolysis of a mixture of the β-lactone (LIV) and pentacarbonyliron(0).

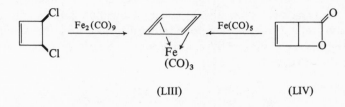

(LIII)   (LIV)

Although flash vacuum-pyrolysis of this complex gives vinylacetylene, mild oxidation with iron(III) or cerium(IV) ions gives free cyclobutadiene as an extremely reactive intermediate which can react *in situ* with a number of compounds. Thus, if the diene is generated in the presence of the appropriate alkyne, the Dewar benzenes (LV; *a–c*) are obtained, which then thermally rearrange to the corresponding benzenoid systems (LVI).

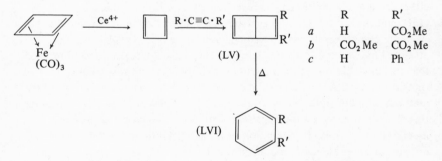

With 2,5-dibromobenzoquinone the diene forms the diketone (LVII) which has been used as an intermediate in the synthesis of the cubane dicarboxylic acid (LVIII).

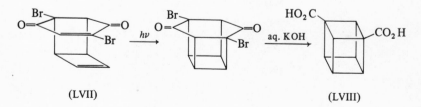

(LVII)   (LVIII)

In the absence of compounds with which the diene can react, oxidation of the iron tricarbonyl complex with cerium(IV) ions gives a number of products including tricyclo-octadiene (LIX), which is presumably formed by dimerisation of the cyclobutadiene.

(LIX)   (LX)

Chemically the iron tricarbonyl complex is similar to ferrocene in that it undergoes electrophilic substitution under mild conditions, and the iron tricarbonyl complexes of a number of mono-substituted cyclobutadienes have been prepared by this type of reaction, e.g. (LX; R = CO.Me, CHO, CH_2Cl and HgCl). By carrying out standard chemical transformations on the functional groups in these complexes a whole variety of substituted cyclo-butadiene complexes have been obtained (cf. p. 178).

Another highly reactive species that forms a stable iron tricarbonyl complex is the diradical trimethylenemethane, which in the ground state is a resonance-stabilised triplet.

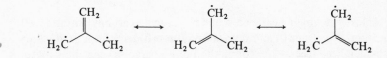

The iron tricarbonyl complex (LXI) is conveniently prepared from 3-chloro-2-(chloromethyl)propene by treatment with either enneacarbonyldi-iron or sodium tetracarbonylferrate (−II), and the trimethylenemethane may be released from the complex by oxidation with cerium(IV) ions or by photolysis. In the absence of suitable trapping agents the free hydrocarbon rapidly reacts with itself to give a number of compounds including the dimer (LXII), but in the presence of activated alkenes or dienes addition products are formed.

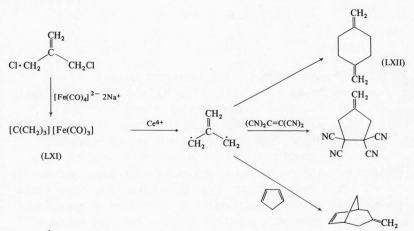

The chemistry of trimethylenemethane and its derivatives and complexes has been reviewed by Weiss (1970).

A number of reactions which involve arynes, nitrenes and carbenes as the reactive intermediates are catalysed by transition-metal ions, e.g. the conversion of the intermediate diazoketone into a ketene in the Wolff rearrangement. On occasions it has been suggested that a complex between the metal ion and the reactive species is involved, and in support of this suggestion stable complexes of various arynes, nitrenes and carbenes have been prepared, e.g. (LXIII) and (LXIV).

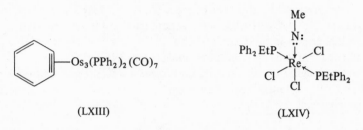

(LXIII)   (LXIV)

Complexes in which a carbene is bonded to pentacarbonylchromium(0), pentacarbonyltungsten(0) and tetracarbonyliron(0) have been particularly well characterised (see p. 266 for method of preparation) and the structures of several have been accurately determined by X-ray diffraction. It appears that in these complexes the carbene acts mainly as a simple electron donor (LXV*b*) and that the electron deficiency of the carbene carbon atom is relieved mainly by π-bonding within the carbene (LXV*c*) rather than by back-donation (LXV*a*) by the metal.

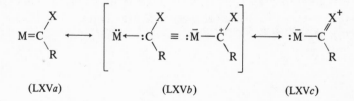

(LXV*a*)   (LXV*b*)   (LXV*c*)

All of the stable complexes of carbenes known at present are of types in which this π-bonding can occur, e.g. (LXVI)–(LXIX), and as a result of the double-bond character of the bond between the carbene carbon and the attached substituents, complexes such as (LXX) are capable of existing as cisoid and transoid forms which can be detected by variable-temperature n.m.r. spectroscopy. In spite of this π-bonding, however, the carbene carbon still bears a substantial positive charge, as revealed by [13]C n.m.r. spectra where the chemical shift of this carbon can be as high as 360 p.p.m. – a value comparable to that of a free carbonium ion. Because of the presence of this positive charge the α-protons in a carbene complex can be removed fairly

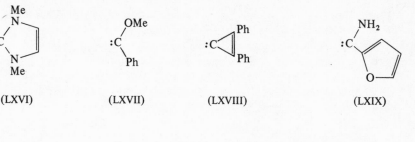

(LXVI)   (LXVII)   (LXVIII)   (LXIX)

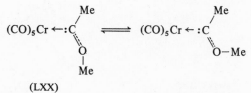

(LXX)

readily, and therefore carbene complexes are relatively strong acids (the pK_a of (LXX) is about 8).

$$\left[M \leftarrow :C\overset{CH_3}{\underset{OMe}{}} \equiv \bar{M}-\overset{+}{C}\overset{CH_3}{\underset{OMe}{}} \right] \overset{-H^+}{\underset{+H^+}{\rightleftharpoons}} \left[\bar{M}-C\overset{CH_2}{\underset{OMe}{}} \leftrightarrow M \leftarrow :C\overset{\bar{C}H_2}{\underset{OMe}{}} \right]$$

Their conjugate bases can be used in organic synthesis for generating C—C bonds, for example, either by direct treatment with reactive electrophilic reagents (aldehydes, acid chlorides, alkylating agents) or by utilising their stability to promote nucleophilic attack on double bonds (see figure 2.1 and the discussion by Casey, 1976).

As indicated by the equations in figure 2.1, useful methods for removing a carbene ligand are oxidation with cerium(IV) ions and treatment with diazomethane.

Carbene complexes are also widely believed to be involved in the metal-catalysed metathesis of alkenes.

$$R^1CH=CHR^2 + R^3CH=CHR^4 \xrightarrow{\text{Catalyst}} R^1CH=CHR^3 + R^2CH=CHR^4$$

In the absence of a catalyst metathesis occurs only at elevated temperatures (in agreement with it being symmetry forbidden according to the Woodward–Hoffman rules), but it can be carried out under relatively mild conditions by using certain homogeneous or heterogeneous catalysts all of which contain a transition-metal component (see reviews by Haines and Leigh, 1975, and Calderon, Ofstead and Judy, 1976). Examples of the catalysed reaction are given below (figure 2.1), but it should be noted that most metatheses involve simple alkenes that contain no other functional groups.

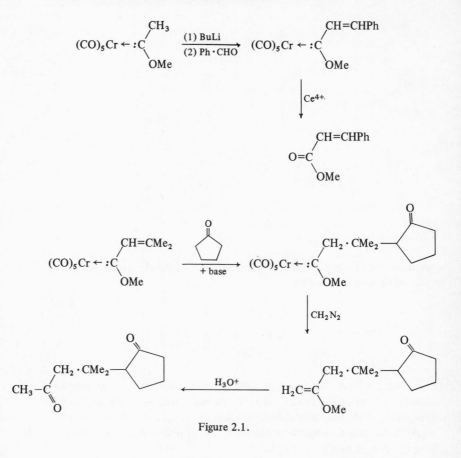

Figure 2.1.

$$H_2C=CH \cdot CN + H_2C=CH \cdot Me \xrightarrow{WO_3/SiO_2/400\,°C} Me \cdot CH=CH \cdot CN + H_2C=CH_2$$

$$\text{(cyclopentene)} + H_2C=CH_2 \xrightarrow{Mo(CO)_6/Al_2O_3/125\,°C} H_2C=CH \cdot (CH_2)_3 \cdot CH=CH_2$$

$$2H_2C=CH \cdot (CH_2)_6 \cdot CO_2 Me \xrightarrow{WCl_6/SnMe_4/70\,°C} MeO_2C \cdot (CH_2)_6 \cdot CH=CH \cdot (CH_2)_6 \cdot CO_2 Me$$
$$+ H_2C=CH_2$$

Although the catalysts used are of two distinct types, i.e. those that incorporate an organometallic alkylating agent and those that do not, both types are currently thought to function by initially generating a carbene complex. With the alkylating type of catalyst this species could be formed *via* a transition-metal alkyl generated by the action of the alkylating agent on one of the other components of the catalyst, for example:

$$W\overset{VI}{-}Cl \xrightarrow{\quad SnMe_4 \quad} W\overset{VI}{-}CH_3 \longrightarrow \underset{\underset{H}{|}}{W}\overset{VI}{\leftarrow}:CH_2$$

With the other type of catalyst it has been tentatively suggested that the carbene complex is formed by the rearrangement of an alkene complex:

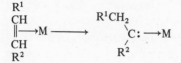

In both cases, however, it is thought that the carbene complex co-ordinates with one of the alkenes that are being subjected to metathesis. A reaction involving a four-membered metallocycle then affords a new carbene complex in which the carbene ligand corresponds to one of the alkylidene groups of the alkene:

$$\underset{R^1CH=\!=\!CHR^2}{\overset{M\leftarrow:CR^5R^6}{\uparrow}} \longrightarrow \underset{R^1CH-CHR^2}{\overset{M-CR^5R^6}{|\quad|}} \longrightarrow \underset{R^1\overset{..}{C}H}{\overset{\overset{CR^5R^6}{\overset{\|}{M\leftarrow|}}}{\uparrow}CHR^2}$$

Alkene exchange is followed by a cycle of reactions which is repeated again and again in a chain process and which results in metathesis.

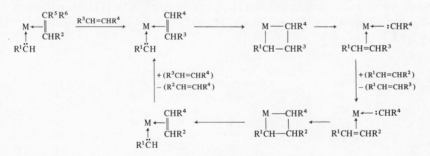

Useful reviews on the preparation and structure of carbene complexes (with emphasis on the isolable ones) are those by Cardin, Cetinkaya and Lappert (1972), and Cotton and Lukehart (1972).

2.7. Displacement of equilibria

Because of the free energy changes which occur when a ligand co-ordinates with a metal, the formation of metal complexes by one or more of the reactants or products of a system at equilibrium invariably causes a displacement in the equilibrium position. Clearly this displacement will be towards the products if these form more stable complexes than the starting materials, and *vice versa*. In practice, complex formation generally causes the equilibrium to be displaced completely to one side, either because the difference between the stabilities of the complexed products and the complexed reactants is sufficiently large, or because one of the metal complexes involved in the equilibrium is insoluble in the reaction medium. If this displacement is to be utilised, in synthetic work for example, it is of course essential to ensure that sufficient metal ions are available for complex formation with the appropriate species to be complete.

Co-ordination with metal ions usually also causes changes in the activation energies of the forward and reverse processes of an equilibrium system, but in this section the discussion concerns only the thermodynamic effect of complex formation and not the kinetics effect. It should be recognised, however, that in fact most of the reactions discussed are catalysed by the metals involved, and that the mechanisms of the catalysed reactions are generally different from those of the uncatalysed ones.

In all the following examples it can be seen that an equilibrium is displaced in favour of the species that has the highest affinity for the metal ions present, either because of the charge or stereochemistry of that species, or because of other factors which increase the stabilities of metal complexes (see §§ 1.10–1.12). The most widely encountered example of equilibrium displacement is the increase in the acidity of many weak organic and inorganic acids which results from co-ordination with a metal ion. This increase arises because the complex formed from the anion of the acid is more stable than that formed from the acid itself. This effect is discussed in more detail in §2.8.

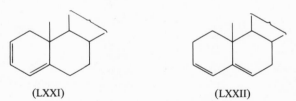

(LXXI)   (LXXII)

Steroidal dienes of the cisoid type (LXXI) are readily isomerised by acids and heat to the more stable transoid $\Delta^{3,5}$ form (LXXII), but unlike the latter type they can displace carbon monoxide from pentacarbonyliron(0) to form stable complexes which have the composition diene–$Fe(CO)_3$. As pentacarbonyliron(0) is one of the many organometallic reagents that can effect the isomerisation of alkenes (see p. 260), if a transoid diene of type (LXXII) is heated with the pentacarbonyl in an inert solvent the iron tricarbonyl complex of the cisoid isomer is obtained. These iron tricarbonyl complexes are decomposed by iron(III) ions to liberate the diene, and hence the overall procedure provides an excellent method for converting a transoid heteroannular diene into the less stable cisoid homoannular isomer. In a similar manner sodium chloroplatinate(II) may be used for preparing $\beta\gamma$-unsaturated ketones from the more stable $\alpha\beta$-unsaturated systems. If 2-methyl-2-penten-4-one, for example, is treated with the sodium salt a good yield of the complex (LXXIII) is formed, from which the free ketone may be liberated by ligand displacement with triphenylphosphine.

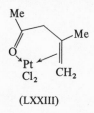

(LXXIII)

One of the many enzymatic transformations of α-amino acids that are dependent upon pyridoxal (LXXIV) or pyridoxamine (LXXV), or both, is the transamination reaction between α-amino acids and α-keto acids (equation 2.5).

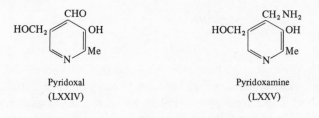

Pyridoxal   Pyridoxamine

(LXXIV)   (LXXV)

$$R \cdot CH(NH_2) \cdot CO_2H + R' \cdot CO \cdot CO_2H \rightleftharpoons R \cdot CO \cdot CO_2H + R' \cdot CH(NH_2) \cdot CO_2H \qquad [2.5]$$

$$R \cdot CH(NH_2) \cdot CO_2H + \text{pyridoxal} \rightleftharpoons R \cdot CO \cdot CO_2H + \text{pyridoxamine} \qquad [2.6]$$

$$R' \cdot CO \cdot CO_2H + \text{pyridoxamine} \rightleftharpoons R' \cdot CH(NH_2) \cdot CO_2H + \text{pyridoxal} \qquad [2.7]$$

Extensive studies which have been summarised by Snell (1968) and more recently by Leussing (1976) indicate that this reaction is the result of sequential coupling of reactions 2.6 and 2.7. Although nearly all the enzymatic reactions do not appear to require metal ions, the model systems that have been investigated are strongly dependent upon them, and the accepted mechanism *in vitro* (figure 2.2 for reaction 2.6) involves the participation of three distinct metal chelates. (The mechanism for reaction 2.7 is the reverse of that of reaction 2.6, except that a second keto acid is one of the two reactants.) With the model systems it is suggested that the main role of the metal is to stabilise the imines formed from pyridoxal and the α-amino acid, and pyridoxamine and the α-keto acid. Unlike the four reactants which are all bidentate ligands, the imines are tridentate and therefore form more stable chelates. In support of this idea quantitative n.m.r. studies of the pyridoxal/alanine/Zn^{2+} and pyridoxamine/pyruvic acid/Zn^{2+} systems have confirmed that the metal

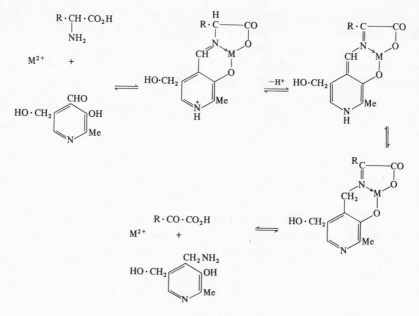

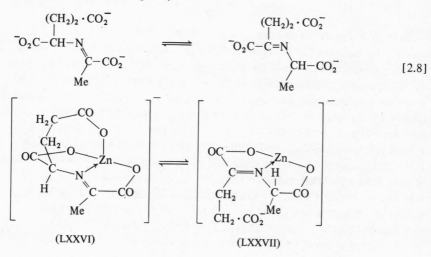

Figure 2.2.

increases the effective concentration of the imines (as chelates) over a wide pH range. One interesting feature which has been revealed by kinetic studies on the formation of imines from glycine and pyridoxal or its analogues, e.g. salicylaldehyde, is that although the reactions are catalysed by certain metal ions such as Zn^{2+} others are ineffective and some, notably Cu^{2+}, actually retard the reactions even though they ultimately displace the equilibrium in favour of the imine (cf. p. 80).

[2.8]

(LXXVI)   (LXXVII)

A further observation from the transamination studies which also demonstrates the effect of chelation is that the equilibrium constant (0.6) for reaction 2.8, i.e. the interconversion of the two Schiff bases formed from pyruvate/glutamate and alaninate/α-ketoglutarate, is about ten times greater than that for the interconversion of their zinc chelates (LXXVI) and (LXXVII). This displacement of the equilibrium position arises because in the chelate of the pyruvate/glutamate Schiff base the carboxylate group in the side-chain is able to co-ordinate with the metal. Although this co-ordination is relatively weak (7-membered ring chelate) it cannot occur in the zinc chelate of the alanate/α-ketoglutarate tautomer on account of the stereochemistry of the imine group. In the two zinc chelates, therefore, one ligand is tridentate and the other is tetradentate, and the latter consequently forms the more stable chelate.

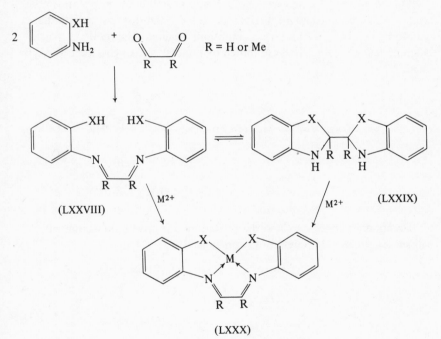

(LXXVIII)

(LXXIX)

(LXXX)

The formation of the di-imines (LXXVIII; X = O) by condensation of 2-aminophenol with either glyoxal or 2,3-butanedione is followed by a rapid intramolecular addition of the —OH groups across the nitrogen–carbon double bonds to give 2,2'-bisbenzoxazolines (LXXIX; X = O). When these are heated in methanol with divalent transition metal ions, however, quantitative yields of the di-imine chelates (LXXX; X = O) are obtained. Kinetic studies with the analogous 2,2'-bisbenzthiazolines (LXXIX; X = S) show that some metals, e.g. Cd^{2+}, rapidly convert the thiazoline directly into the chelate

(LXXX; X = S), but others such as Zn^{2+} react very slowly with the thiazoline and function largely by irreversibly displacing the equilibrium between the thiazoline and the di-mine by chelating with the latter.

Another example in which the thermodynamic effect is utilised in organic synthesis is the use of methylmagnesium carbonate (MMC) to effect the sequence of reactions 2.9, i.e. the reverse of the usual mechanism by which carboxylic acids are decarboxylated.

$$L \cdot CRR' \cdot H \longrightarrow L \cdot \bar{C}RR' \xrightarrow{CO_2} L \cdot CRR' \cdot CO_2^- \xrightarrow{H^+} L \cdot CRR' \cdot CO_2H$$

L = a co-ordinating group, e.g. $\diagdown$CO, NO_2 [2.9]

The methylmagnesium carbonate is first prepared in dimethylformamide from carbon dioxide and magnesium methoxide, and then a substrate that can form a carbanion under basic conditions is added. The resultant magnesium complex is carefully decomposed by mineral acid to give the carboxylic acid.

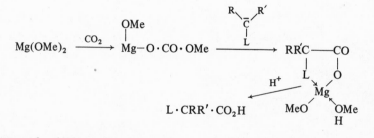

Compounds which can be carboxylated by this procedure include nitro-alkanes, ketones, and polyhydroxy phenols.

$$Me \cdot NO_2 \longrightarrow HO_2C \cdot CH_2 \cdot NO_2$$

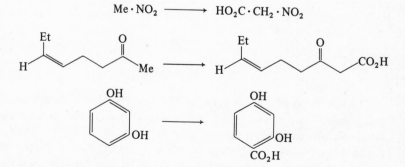

When ketones and related compounds that have at least two α-hydrogen atoms are used, the intermediate magnesium salts may be alkylated *in situ*, thus permitting additional synthetic application of the reagent.

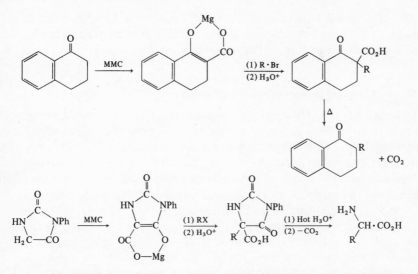

It is currently thought that these *in vitro* carboxylations with methyl-magnesium carbonate are directly analogous to those biological carboxylations that involve metallo-enzymes, e.g. the manganese(II) enzyme which is associated with the carboxylation of pyruvic acid:

$$\text{enzyme–biotin–}CO_2 + \text{pyruvate} \rightarrow \text{enzyme–biotin} + \text{oxaloacetate}.$$

One final point concerning the displacement of mobile equilibria by metal ions is that on a number of occasions it has been stated that an example of this effect is the conversion of acetylacetone into metal acetylacetonates. As this diketone normally contains about 80 and 20 per cent of the keto and enol tautomers respectively, it is suggested that the metal ion displaces the equilibrium between the tautomers by completely converting the diketo form into the co-ordinated enol form. It must be pointed out, however, that in the metal acetylacetonates the ligand is not the uncharged enol but the resonance stabilised anion which is generated from the diketo form – as well as from the enol form – by loss of a proton. In the formation of the acetylacetonates, therefore, it is the equilibria involving the ionisations of the keto and the enol tautomers that are displaced, not the equilibrium between these two

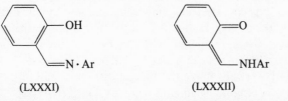

(LXXXI)	(LXXXII)

tautomers. A more appropriate example to illustrate this particular effect is the reaction of hexa-aquocobalt(II) chloride with *N*-arylsalicylaldimines

(LXXXI), which affords complex cations of the composition $[(\text{ligand})_3\text{Co}]^{2+}$ in which the ligand is the tautomeric form (LXXXII) of the starting imine, and which is chelated through the oxygen and nitrogen atoms.

2.8. Increased acidity of co-ordinated acids

The co-ordination of an organic or inorganic acid (HL) with a metal ion results in an increase in the acidity of the acid if the acid is co-ordinated through the atom that bears the negative charge in the conjugate base (L^-) produced by ionisation. One of the factors that are responsible for this is that co-ordination of the acid with a positively charged species causes additional polarisation of the $H-L$ bond and hence lowers the heat of ionisation of the acid. Another and probably more important factor is that upon ionisation the negative charge of the conjugate base and the positive charge of the metal ion partly neutralise each other, and the overall loss of ionic charge results in a decrease in the number of solvent molecules required for solvation. Co-ordination therefore increases the value of ΔS of solvation, and a combination of this increase and the decrease in ΔH of ionisation causes $-\Delta G$ to be greater for equation 2.11 than for equation 2.10.

$$H-L \rightleftharpoons H^+ + L^- \qquad\qquad [2.10]$$

$$H-L\!: \rightarrow M^{n+} \rightleftharpoons H^+ + \ddot{L}-M^{(n-1)+} \qquad\qquad [2.11]$$

In some cases stable metal complexes of the deprotonated ligand may be isolated, but quantitatively the increase in acidity is revealed by titration curves which on analysis afford the pK_a value of the co-ordinated acid. The rates of those reactions that involve conjugate base formation in the rate determining step, e.g. hydrogen–deuterium exchange, can also give a quantitative measure of the increase in acidity.

Most of the early quantitative measurements were concerned with the acidity of aquo and ammine complexes, some of which are quite strong acids. For the ion $[\text{Hg}(\text{H}_2\text{O})_2]^{2+}$, for example, $pK_a = 3.7$, which is about the same as that of monochloroacetic acid. The acidities of co-ordinated water and ammonia molecules are dependent upon the stereochemistry of the complex and the charge and electronic configuration of the metal ion (see table 2.1), but as expected the small, highly charged ions are the most effective at increasing the acidity. It should be noted that although the oxonium ion is more acidic than the ammonium ion by a factor of about 10^{11}, the acidity of a co-ordinated water molecule in a highly charged complex such as $[\text{Pt}(\text{NH}_3)_5(\text{H}_2\text{O})]^{4+}$ is only about 10^3 times greater than that of the ammonia molecules in the corresponding ammine complex, i.e. $[\text{Pt}(\text{NH}_3)_6]^{4+}$. Presumably this reflects greater charge dispersal in the conjugate base of the ammine complex by means of $(d-p)_\pi$ bonding between the metal and the

Table 2.1. *Acid ionisation constants of co-ordinated water, ammonia and 1,2-diaminoethane*

Complex[a]	pK_a (25 °C)
$[Ti(H_2O)_n]^{3+}$	2.2
$[V(H_2O)_n]^{3+}$	2.8
$[Cr(H_2O)_n]^{3+}$	3.8
$[Mn(H_2O)_n]^{2+}$	10.6
$[Fe(H_2O)_n]^{3+}$	2.2
$[Fe(H_2O)_n]^{2+}$	9.5
$[Co(H_2O)_n]^{3+}$	0.7
$[Co(H_2O)_n]^{2+}$	8.9
$[Ni(H_2O)_n]^{2+}$	10.6
$[Cu(H_2O)_n]^{2+}$	6.8
$[Ru(NH_3)_6]^{3+}$	12.4
$[Pt(NH_3)_6]^{4+}$	7.2
$[Pt(NH_3)_5H_2O]^{4+}$	4.0[b]
$[Pt(NH_3)_5Cl]^{4+}$	8.4
cis-$[Pt(NH_3)_4Cl_2]^{2+}$	9.7
trans-$[Pt(NH_3)_4Cl_2]^{2+}$	11.3
$[Pt(en)_3]^{4+}$	5.5
$[Co(en)_3]^{3+}$	15.2

[a]With most hydrated metal ions the exact number of co-ordinated water molecules has not been determined. With all of the ions listed, however, this number may be assumed to be six.
[b]The pK_a of the co-ordinated H_2O.

negatively charged nitrogen. For a more detailed discussion on the acidities of aquo and ammine complexes, the reader is referred to the discussion by Basolo and Pearson (1967).

Chelated ligands are more acidic than the analogous complexed ones, and consequently the order of acidity $[Pt(en)_3]^{4+} > [Pt(en)_2(NH_3)_2]^{4+} > [Pt(en)(NH_3)_4]^{4+} > [Pt(NH_3)_6]^{4+}$ is observed even though $H_2N.(CH)_2.^+NH_3$ is a weaker acid than $^+NH_4$. The cation $[Co(en)_3]^{3+}$ exchanges hydrogen for deuterium in D_2O faster than does $[Co(NH_3)_6]^{3+}$, and a number of stable deprotonated ethylenediamine complexes with ions such as Ir^{3+}, Pd^{2+}, Os^{4+} and Rh^{3+} have been prepared. For example, deprotonation of the salt (LXXXIII; R = H) by ammonia affords (LXXXIV; R = H).

The ethylenediamine ligands in the salt (LXXXV; R = H) are less acidic than those in (LXXXIII; R = H), and require potassium amide in liquid ammonia for deprotonation. Similarly the salt (LXXXV; R = Me) undergoes hydrogen–deuterium exchange at the secondary amine groups about 300 times slower in D_2O than the salt (LXXXIII; R = Me). The greater acidity of these bipyridyl complexes is due to the ability of this ligand to delocalise the negative charge of the conjugate base by π-bonding through the metal.

Chelating alcohols are more acidic than their monodentate analogues, and thus the co-ordinated —OH groups in chelates such as $[Co(en)_2(H_2N.CH_2.$

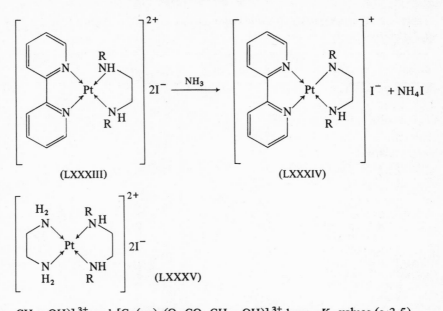

(LXXXIII)   (LXXXIV)

(LXXXV)

$CH_2 . OH)]^{3+}$ and $[Co(en)_2(O . CO . CH_2 . OH)]^{3+}$ have pK_a values (~3.5) approximately two units lower than those in the analogous systems that contain monodentate alcohols. At a pH value of 8.5, N-(2-hydroxyethyl)ethylenediamine is deprotonated in the presence of cobalt(III) ions to give a chelate in which both nitrogens and the negatively charged oxygen are co-ordinated with the metal; 2-hydroxyethylamine behaves similarly in the presence of copper(II) ions. It is possible, of course, that in the last two examples the un-ionised hydroxyl groups in the initially-formed complexes are not co-ordinated with the metal, and that the ionisations that occur at comparatively low pH values are the result of the metal displacing the equilibrium between the hydroxyl group and the alkoxide anion by combining with the latter. A situation similar to this could possibly be present in the conversion of π-complexes of alkenes with palladium chloride into π-allyl complexes. This conversion, which involves deprotonation at the allylic position, can be accomplished at room temperature even by weak bases such as sodium hydrogen carbonate (for convenience the reaction is represented in the scheme as involving the monomeric species).

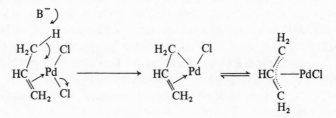

The acidity of amides is also substantially higher when they are chelated with a metal ion rather then simply complexed. Although the mercury(II) compounds $(R' . CO . NR)_2 Hg$ of primary and secondary amides are easily prepared by treatment of the amide with mercury(II) oxide and are suitable derivatives for the characterisation of amides, proton loss from a complexed amide usually takes place only under fairly basic conditions. In contrast, the chelates (LXXXVI; M = Ni, Cu and Co) and (LXXXVII) are readily formed under mildly basic conditions from the appropriate metal(II) ion and the parent diamide.

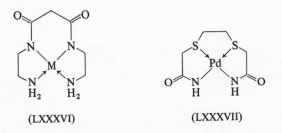

(LXXXVI)   (LXXXVII)

Most of the work on the acidity of chelated amides has been concerned with model systems for the interaction between transition-metal ions and proteins, and hence the majority of the amides used have been simple di-, tri- and tetrapeptides. As expected, potentiometric titrations and aqueous (D_2O) infrared and proton n.m.r. spectroscopy have revealed that these types of compounds exhibit ionisations of their amide groups at comparatively low pH values when they are co-ordinated with transition-metal ions. Typically, in the presence of an equivalent amount of copper(II) ions, glycylglycine loses both protons from the carboxyl and amide groups below pH = 7.9; the pK_a value of the co-ordinated amide group in (LXXXVIII) is 5 compared with 17 in the free ligand.

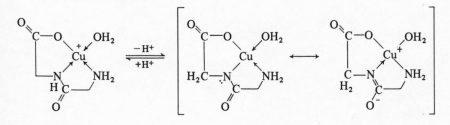

(LXXXVIII)

An interesting feature of the ionisation of the two amide groups of diglycylglycine is that when this peptide is co-ordinated with Ni^{2+}, the second acid dissociation constant $(10^{-7.7})$ is actually higher than the first $(10^{-8.8})$. This unusual situation arises because of differences between the electronic states and the co-ordination numbers of the two species (LXXXIX) and (XC) which undergo ionisation.

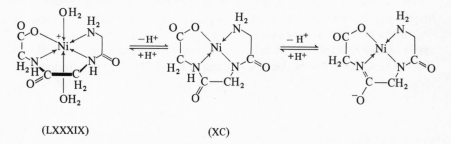

(LXXXIX)   (XC)

The preparations and structures of many transition-metal complexes of simple peptides have been reviewed by Freeman (1967).

In all the preceding examples, the negative charge formed by proton removal was located specifically on one of the atoms in the ligand directly co-ordinated with the metal. Clearly the negative charge of a conjugate base will also be stabilised when the proton is removed from the ligand at a distance from the donor atom if the negative charge is resonance stabilised and the donor atom participates directly in the resonance stabilisation. As in the simpler systems, the increase in stability of the conjugate base is reflected not only by the greater acidity of the co-ordinated ligand, but also by the

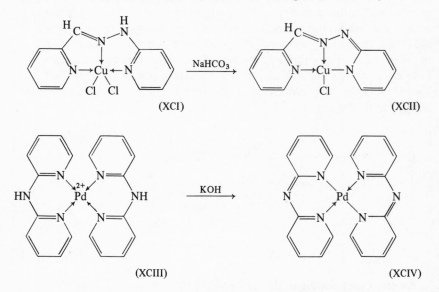

higher rates of those reactions of the ligand that involve the conjugate base in the rate-determining step. Stable complexes of deprotonated ligands have been obtained in a number of cases, particularly those in which the acidities of hydroxyl and amino groups in the 2- and 4-positions of a pyridine ring have been increased by co-ordination of the pyridine nitrogen. The hydrazone (XCI), for example, has been deprotonated under mild conditions by sodium hydrogen carbonate to give the chelate (XCII). The chelate (XCIV) has similarly been obtained by treatment of the cation (XCIII) with potassium hydroxide in methanol.

$$\overset{|}{\underset{/}{\overset{\backslash}{C}}}-\overset{|}{C}=X \to M^{n+} \longleftrightarrow \overset{\backslash}{\underset{/}{C}}=\overset{|}{C}-\ddot{X}-M^{(n-1)+} \qquad [2.12]$$

Of particular interest to organic chemists is the additional stabilisation bestowed upon imine and enol anions by co-ordination of their nitrogen and oxygen atoms (equation 2.12; X = NR or O), for these co-ordinated species play important roles in many metal-catalysed reactions of imines and carbonyl compounds. Some of these are briefly mentioned below.

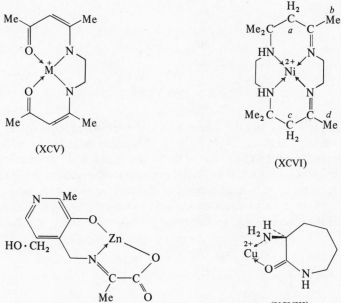

(XCV)

(XCVI)

(XCVII)

(XCVIII)

The rates of α-deuteriation of co-ordinated ketones and derived imines are much higher than those of the unco-ordinated ligands, and when the cation (XCV; M = Co(NH$_3$)$_2$) is heated in basic D$_2$O at 100 °C for 10 minutes the acetyl methyl groups are completely deuteriated. Rapid and selective catalysis of H–D exchange occurs in D$_2$O at room temperature at positions *a, b, c* and *d* of the macrocyclic chelate (XCVI) and at the pyruvate methyl group of the complexed Schiff base (XCVII). Stabilisation of the carbanion formed from (XCVIII) also explains the copper-catalysed racemisation of *S*-3-amino hexahydro-2H-azepin-2-one.

The higher acidity of the α-hydrogen atoms of an *O*-co-ordinated acetyl group has been revealed by kinetic studies of various metal-catalysed reactions of ketones. A spectrophotometric investigation has shown that in acid solution the interconversion of the keto and enol tautomers of acetylacetone is catalysed by divalent transition metal ions, the order of effectiveness being Co > Ni > Zn > Mn > Cd. As the interconversion is not acid-catalysed but proceeds simply by ionisation followed by a recombination step (figure 2.3), this ob-

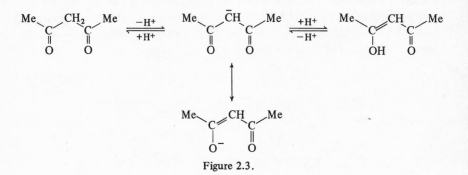

Figure 2.3.

servation indicates that the diketo form is deprotonated more readily when it is co-ordinated as in (XCIX; M = Co^{2+}). This is also indicated by stop-flow kinetic studies of the formation of the cation $[Cu(acac)]^+$ from acetylacetone and Cu^{2+} in aqueous solution. Relevant to these observations is the fact that in addition to the very large number of chelates formed by the anion (acac) of acetylacetone, a few chelates are known in which the ligand is the neutral diketo form of the diketone, e.g. $[Ni(acacH)_3](ClO_4)_2$ and (XCIX; M = $SnCl_4$).

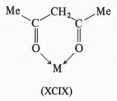

(XCIX)

The catalytic effect of metal ions on keto–enol tautomerism is also manifested in the rearrangement of α-ketohemimercaptals into α-hydroxythiolesters. The rearrangement in dimethylformamide with *N*-methylpyrrolidine as a basic catalyst is strongly promoted by a variety of metal ions, notably Mg^{2+}.

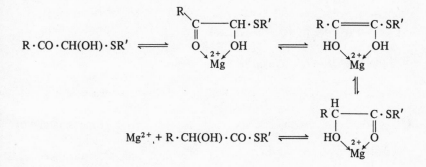

As the α-ketohemimercaptals may be obtained by equilibrating α-keto-aldehydes with thiols, this rearrangement is of interest in connection with the biochemical conversion of methyl glyoxal into lactic acid.

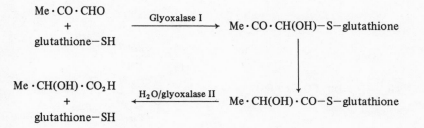

The brominations of ethyl acetoacetate and 2-ethoxycarbonylcyclo-pentanone in acid solution are also catalysed by transition-metal ions, and the reactions are kinetically of zero order with respect to the halogen. This is consistent with a mechanism in which loss of a proton from the co-ordinated diketo form is the rate-determining step (figure 2.4).

The bromination and iodination of pyruvic acid as well as the condensation of this acid with aldehydes are similarly catalysed by metal ions, and here again the metal probably functions by accelerating the formation of the reactive species, i.e. the carbanion (see figure 2.5).

Chelation by the pyridine nitrogen and the carbonyl oxygen also explains the large catalytic effect that Cu^{2+}, Ni^{2+} and Zn^{2+} have on the iodination of 2-acetylpyridine.

The final example of resonance-stabilised conjugate bases gaining additional stability as the result of co-ordination is provided by cobalt(III) cations of the type $[(NH_3)_5Co \leftarrow NC.CH_2.R]^{3+}$; in these species the acidity of the $-CH_2-$ group is enhanced by *ca.* 10^6 because the negative charge of the conjugate base is stabilised by the metal.

$$Co^{3+} \leftarrow N \equiv C - \bar{C}H \cdot R \longleftrightarrow Co^{2+} - \ddot{N} = C = CH \cdot R$$

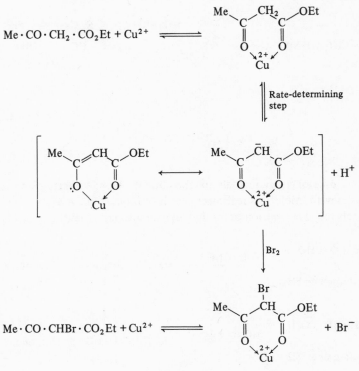

Figure 2.4.

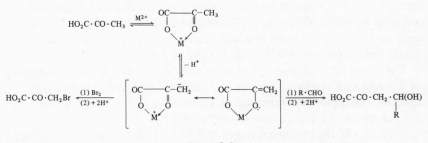

Figure 2.5.

2.9. Metal-catalysed decarboxylation of β-keto acids

The increase in the stability of enolate anions which results from co-ordination of the oxygen atom is of particular importance in the decarboxylation of certain β-keto acids. In solution the decarboxylation of a β-keto acid proceeds not only *via* the carboxylate anion (equation 2.13) as is usually the case with carboxylic acids, but also directly through a six-membered ring

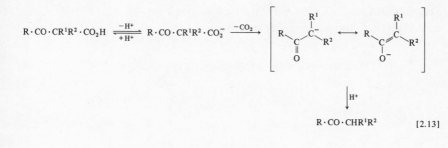

$$R \cdot CO \cdot CR^1R^2 \cdot CO_2H \; \underset{+H^+}{\overset{-H^+}{\rightleftharpoons}} \; R \cdot CO \cdot CR^1R^2 \cdot CO_2^- \xrightarrow{-CO_2}$$

[2.13]

$$R \cdot CO \cdot CHR^1R^2$$

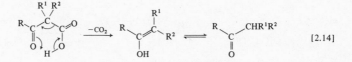

[2.14]

transition state (equation 2.14). Although the latter process is usually the faster of the two, the rate of the former is considerably increased if the ketonic oxygen atom is co-ordinated with a metal ion because of the additional stabilisation bestowed upon the enolate anion formed in the rate-determining step. For co-ordination to occur to any extent the carbonyl group must have an affinity for metal ions which is at least comparable with that of the ionised carboxylate group. This condition is only fulfilled when the carbonyl group is adjacent to another co-ordinating group (L), i.e. when the carbonyl group forms part of a chelating system as shown in the general structure (C).

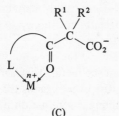

(C)

This metal-ion catalysis has been investigated by many groups of workers, mainly because the activation of several enzymes that are responsible for the biochemical decarboxylation of β-keto acids is dependent on, or in some cases is enhanced by, specific metal ions, notably Mn^{2+} and Mg^{2+}. α,α-Dimethyl-oxaloacetic acid has often been used in these investigations because, unlike the unsubstituted acid, it cannot exhibit keto–enol tautomerism, and this simplifies kinetic studies of the decarboxylation. Figure 2.6 shows the accepted mechanism for the metal-catalysed decarboxylation of this compound in aqueous solution, and it may be safely assumed that related keto-acids, e.g. acetonedicarboxylic acid, are decarboxylated by an identical process. In accord with this mechanism the decarboxylation of the half ethyl ester

$EtO_2C.CO.CMe_2.CO_2H$ is not catalysed by metal ions because of the very weak chelating properties of α-keto esters.

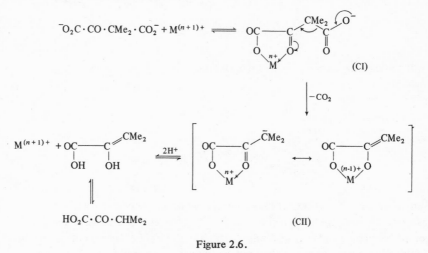

Figure 2.6.

The catalytic activity of any metal ion in the decarboxylation of α,α-dimethyloxaloacetic acid passes through a maximum at a specific metal ion : keto acid ratio, and then progressively decreases as the metal-ion concentration increases; high concentrations of the metal actually inhibit the reaction. This is because the unco-ordinated carboxylate anion in the chelate (CI) is essential for the decarboxylation, and this group cannot remain unco-ordinated once the metal requirements of the adjacent chelating system have been satisfied.

The efficiency of a particular metal ion in catalysing the decarboxylation may be altered by changing the effective charge on the metal. This is most conveniently done by adding to the reaction mixture other ligands that can displace the water molecules occupying the remaining co-ordination positions of the metal in the chelate (CI). Ligands such as piperidine, methylamine and acetate ions will do this, and being stronger σ-donors than water they reduce both the effective change on the metal and the stability of the co-ordinated enolate anion (CII) formed in the rate-determining step. The catalytic decomposition is therefore retarded by these compounds, but conversely it is accelerated by ligands that can participate in π-bonding with the metal and help to stabilise the co-ordinated anion by delocalising the negative change. Pyridine, 2,2'-dipyridyl and 1,10-phenanthroline fall into this category, and in the phenanthroline–manganese(II) catalysed decarboxylation of α,α-dimethyloxaloacetic acid the catalytic activity of three of the cations present decreases in the order $[Mn(phen)_2]^{2+} > [Mn(phen)]^{2+} > [Mn(H_2O)_n]^{2+}$

In the catalysed decarboxylation of oxaloacetic acid the effectiveness of various divalent metal ions follow the order $Ca^{2+} < Mn^{2+} < Co^{2+} < Zn^{2+} < Ni^{2+} < Cu^{2+}$. Although there is little correlation between the overall rate constants of the catalysed reactions and the stability constants of the corresponding 1 : 1 metal complexes of oxaloacetic acid, there is almost a linear correlation between the former and the stability constants of the oxalate complexes. Similarly in the metal–ion catalysed decarboxylation of acetonedicarboxylic acid there is a very good correlation with the stability constants of the corresponding malonate complexes. It has been suggested that these correlations show that the transition states of the decarboxylations resemble the initial products, i.e. (CIII) and (CIV), more than they do the starting chelates.

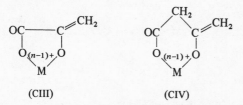

(CIII)   (CIV)

With iron(III) ions the decarboxylation of α,α-dimethyloxaloacetic acid is accompanied by striking colour changes. On dissolving the acid in an almost colourless solution of iron(III) ammonium sulphate the initial bright yellow colour due to a chelate of type (CI) progressively changes to green, blue and deep blue as the decarboxylation takes place and the chelate (CII) is formed. The colour then gradually fades and the solution is finally almost colourless.

2.10. Stabilisation of carboxylate carbanions

When judged as an electron-releasing or electron-withdrawing substituent the carboxylate anion ($-CO_2^-$) is found to be very much a borderline case. The ionisation constants of substituted benzoic acids reveal that in the *meta-* or *para*-positions the group is weakly electron-releasing, but in the hydrolysis of aromatic esters of the types $Ar \cdot CO_2Et$ and $Et \cdot CO_2Ar$ it is definitely electron-withdrawing. If the negative charge on the anion is first reduced to some extent by salt formation with a metal ion, the carboxylate group will stabilise an adjacent negatively-changed carbon atom.

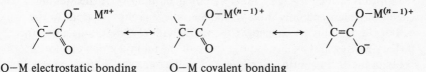

O$-$M electrostatic bonding   O$-$M covalent bonding

Substantial stabilisation occurs even with carboxylate salts of Group Ia and IIa metals in spite of the high ionic character of the oxygen–metal bonds.

This is revealed by the fact that if sodium or potassium propionate is refluxed in basic D_2O, the α-hydrogen atoms are replaced by deuterium.

$$Me \cdot CH_2 \cdot CO_2Na \xrightleftharpoons{DO^-} Me \cdot \bar{C}H \cdot CO_2Na \xrightleftharpoons{D_2O} Me \cdot CHD \cdot CO_2Na \qquad [2.15]$$

Lithium salts of straight-chain carboxylic acids are metallated exclusively at the α-position by treatment with lithium di-isopropylamide (figure 2.7), and the resultant α-lithium compounds are very useful synthetic intermediates.

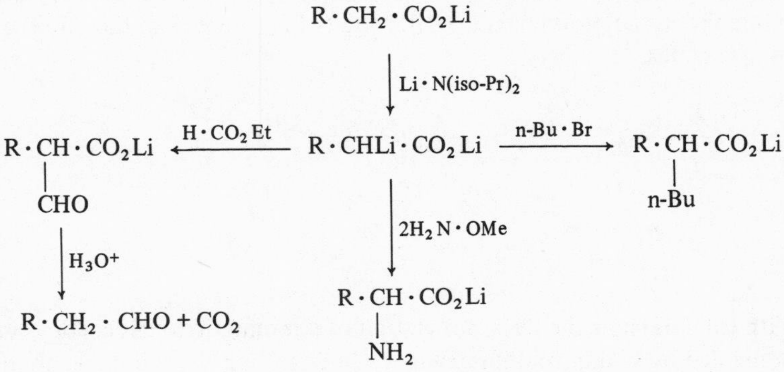

$$R \cdot CH_2 \cdot CO_2Li$$

$$\downarrow Li \cdot N(iso\text{-}Pr)_2$$

$$R \cdot CH \cdot CO_2Li \xleftarrow{H \cdot CO_2Et} R \cdot CHLi \cdot CO_2Li \xrightarrow{n\text{-}Bu \cdot Br} R \cdot CH \cdot CO_2Li$$

CHO   n-Bu

$$\downarrow H_3O^+ \qquad\qquad \downarrow 2H_2N \cdot OMe$$

$$R \cdot CH_2 \cdot CHO + CO_2 \qquad\qquad R \cdot CH \cdot CO_2Li$$

NH$_2$

Figure 2.7.

Carbanion formation at the α-position of carboxylate salts has been extensively studied by examining base-catalysed hydrogen–deuterium exchange in metal chelates of simple α-amino acids and peptides, and related compounds such as EDTA. As expected, it has been found that carbanion formation proceeds increasingly readily as the negative charge on the $-CO_2^-$ group becomes less delocalised and more closely associated with the oxygen–metal bond. With EDTA chelates of Group IIa metal ions the order of ease of carbanion formation is $Mg^{2+} > Ca^{2+} > Sr^{2+} > Ba^{2+}$ which reflects the size : charge ratio of the metal ions, while with transition metals the observed order $Cu^{2+} > Ni^{2+} > Co^{2+} > Zn^{2+} > Pb^{2+}$ may be rationalised in terms of the ligand field and the degree of covalent character of the oxygen–metal bond. In accord with the proposed mechanism of the exchange (cf. equation 2.15), the protons of the $-\overset{|}{N} . (CH_2)_2 . \overset{|}{N}-$group in all the EDTA chelates are chemically inert and only those of the methylene groups adjacent to the carboxylate groups are exchanged. Similarly with chelates of α-alanine the tertiary hydrogen is exchanged but those of the methyl group are not. Increasing the charge on the metal ion causes the expected increase in the lability of the α-hydrogens, and hydrogen–deuterium exchange proceeds more rapidly in the alanine residue of the cationic species $[Co(III)en_2(L\text{-ala})]^{2+}$ than in those of the neutral chelate $[Co(III)(L\text{-ala})_3]^0$.

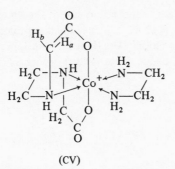

(CV)

In chelates that contain a pair of diastereotopic hydrogens in the α-position to the carboxylate group, deuteriation can be stereoselective. With the anion (CV), for example, H_a is replaced by deuterium faster than H_b, because for steric reasons the former hydrogen is abstracted as a proton more readily than the latter.

Because of the additional resonance stabilisation, the carbanion formed from a chelate of a Schiff base of an α-amino acid is considerably more stable than that derived from the corresponding chelate of that α-amino acid.

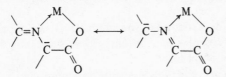

Deuteriation and racemisation at the α-position of the *N*-salicylidene chelates (CVI) consequently proceed far more rapidly than in the corresponding $[Cu(II)(\alpha\text{-amino acid})_2]^0$ chelates. In basic solution carbanion formation occurs preferentially at C_4 in dipeptide complexes of type (CVII) rather than at C_2, and under the minimum conditions required completely to racemise the (L)-alanylglycine chelate (CVII; R^1 = Me, R^2 = H) the glycyl(L)-alanine chelate (CVII; R^1 = H, R^2 = Me) is optically stable. Base-catalysed deuteriation of the corresponding glycylglycine chelate, followed by degradation with hydrochloric acid gives the specifically dideuteriated peptide, $H_2N.CD_2.CO.NH.CH_2.CO_2H$.

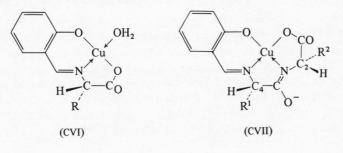

(CVI)   (CVII)

This preferential carbanion formation at the amino end of the dipeptide derivative contrasts sharply with what is observed when the dipeptides themselves are chelated to cobalt(III) in the anions $[Co(III)(dipeptide)_2]^-$, for in these systems proton abstraction (and hence deuteriation) occurs specifically

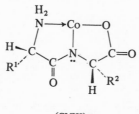

(CVIII)

at C_2 (see partial structure (CVIII)), presumably as a natural result of the order of stability:

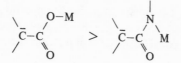

During the base-catalysed deuteriation of the anion $[Co(III)(gly-L-ala)_2]^-$ the circular dichroism of the complex remains unchanged, suggesting that the deuteriation is completely stereospecific.

Attempts to utilise the nucleophilic properties of carbanions derived from chelates of α-amino acids and peptides have met with varying degrees of success. It is clear, however, that by use of suitable systems it should ultimately become possible to synthesise a variety of α-amino acids and peptides by incorporating a glycine residue, for example, into a metal chelate and then generating the α-carbanion and using it to attack electrophilic substrates such as alkyl halides and carbonyl compounds. The following discussion briefly describes how this technique has been applied to fairly simple systems.

Under mildly basic conditions at room temperature the glycine chelate (CVI; R = H) reacts with benzaldehyde, acetaldehyde and formaldehyde to give the analogous chelates of β-phenylserine, threonine and serine respectively. These chelates are decomposed by acid to give the free α-amino acid, and except in the case of serine the yields are quite good. The copper(II) chelate (CIX) of N-pyruvylglycine reacts similarly with a variety of aldehydes, and with 2,3-O-isopropylidene-D-glyceraldehyde a 70 per cent yield of the 2-amino-aldopentonic acid derivative (CX) is obtained.

In the so-called Akabori reaction, bis(glycinato)copper(II), acetaldehyde and aqueous sodium carbonate are heated together at 50 °C for 1 hour to give the bis(oxazolidine-carboxylate) chelate (CXII), which on degradation with hydrogen sulphide gives a 64 per cent overall yield of racemic threonine

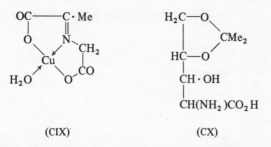

(CIX)   (CX)

and allothreonine with the former as the major product. Each oxazolidine
ring in the chelate (CXII) is presumably formed by the α-carbanion of the
Schiff base (CXI) reacting with acetaldehyde to give an alkoxide anion which
attacks the co-ordinated carbon–nitrogen double bond. The participation of
a chelated Schiff base also explains the formation of serylglycine (70 per
cent yield) when glycylglycine, formaldehyde, copper sulphate and aqueous
sodium carbonate are heated together at 100 °C for 1 hour. Analogous to
this last reaction is the condensation of the chelate (CVII; $R^1 = R^2 = H$) with
benzaldehyde, which as expected takes place at C_4 to give a chelate from
which (β-phenylseryl)glycine may be obtained by degradation.

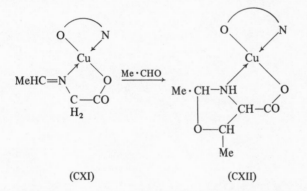

(CXI)   (CXII)

   In the condensations outlined above, the amino acid or peptide was in the
form of a co-ordinated Schiff base from which the reactive carbanion was
generated fairly easily. In addition to these *in vitro* reactions, there are a
number of enzymatic processes which involve the α-position of an α-amino
acid and which are readily rationalised in terms of carbanion intermediates.
These processes, which include racemisation, transamination (see p. 101),
and the interconversion of glycine with serine or threonine, are all believed
to be initiated by formation of a Schiff base with pyridoxal or one of its
derivations. Although most of the enzymes responsible for these transformations have no known requirement for metal ions, the model systems that have
been studied are very dependent on them, and it is evident from these model
systems as well as from the preceding discussion that the incorporation of an

α-amino acid into a chelate of type (CXIII) greatly facilitates removal of the α-hydrogen as a proton.

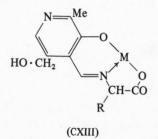

(CXIII)

It should be noted, however, that under basic conditions alkylation of the cation $[\text{Co(III)(en)}_2\text{Gly}]^{2+}$ by benzyl bromide and methyl iodide followed by acid degradation affords phenylalanine (10 per cent) and alanine (20 per cent) respectively on degradation. This clearly shows that it is not essential to form a Schiff base in order to carry out reactions at the α-position of an amino acid.

3   Metal complexes in substitutions

3.1.   Introduction

In a nucleophilic substitution at a saturated carbon there are four essential
components, the nucleophile, the substrate, the displaced leaving group, and
substituted product. Both the nucleophile (Nu:) and the displaced leaving
group (X:) possess at least one lone pair of electrons, and can therefore act
as donor ligands for a metal ion. The substrate and the product may also act
as ligands if there is a lone pair of electrons on the groups X and Nu when
these groups are bonded to a saturated carbon, e.g. when X = Br and Nu = OH.
As a result, metal ions can affect the kinetics and thermodynamics of a
nucleophilic substitution by co-ordinating with one or more of the species
involved in the reaction. One useful result of this co-ordination is the increase
in the rate of reaction that is associated with a decrease in the activation
energy. In order to explain how this may arise the reader is reminded that
the mechanism for a nucleophilic substitution at a saturated carbon may be
considered to lie between two extreme pathways, i.e. the so-called S_N1 and
S_N2 mechanisms. In the former the bond between the saturated carbon and
the leaving group is broken before the bond to the nucleophile is formed
(equation 3.1), while in the latter both processes are completely concerted
(equation 3.2).

$$-\overset{|}{\underset{|}{C}}-X \xrightarrow{-X:} \overset{|}{\underset{/\backslash}{C^+}} \xrightarrow{+Nu:} -\overset{|}{\underset{|}{C}}-Nu \qquad [3.1]$$

$$Nu: \,\,\, \overset{\backslash}{\underset{/}{C}}-X \longrightarrow Nu-\overset{/}{\underset{\backslash}{C}} + X: \qquad [3.2]$$

When the substrate co-ordinates with a metal ion by means of a lone pair
of electrons on the group X, the co-ordination increases the polarisation of

the $C-X$ bond towards the group X and facilitates the departure of that group with the pair of bonding electrons. Less energy is therefore required for bond breaking. Furthermore, the loss of entropy associated with solvation of a co-ordinated leaving group is much lower than the loss associated with solvation of a free leaving group, because of the effective reduction of electron density on the leaving group by the metal. As this factor usually more than compensates for the attendant decrease in the enthalpy of solvation, the thermodynamic stability of the leaving group is increased by co-ordination to the metal. This is reflected in the transition state of the substitution, and the activation energy is consequently lowered.

From stereochemical and kinetic studies it appears that most metal-ion promoted nucleophilic substitutions have a high degree of $S_N 2$ character, i.e. they involve both the 'pull' of the metal on the leaving group and the 'push' of the incoming nucleophile in a concerted process. Although bond breaking may be substantially more advanced than bond formation in the transition state, the substitution only proceeds by a genuine two-step $S_N 1$ mechanism when conditions are extremely favourable for carbonium-ion formation, as in reactions that involve strong Lewis acids, such as aluminium trichloride, and tertiary or benzylic halides. The majority of silver-ion catalysed nucleophilic substitutions, for example, seem to occur by a concerted process and do not involve carbonium-ion formation. One stereochemical consequence of this situation is that inversion of configuration, rather than racemisation, is commonly observed even with systems where the substrate would be expected to acquire substantial carbonium-ion character in the transition state.

In nucleophilic substitutions where the nucleophile is used in conjunction with a metal cation, e.g. as a salt, the activation energy of the reaction may be effectively lowered by deliberate use of a metal ion that has a low affinity for the nucleophile but a high affinity for the leaving group. Thus in the preparation of an alkyl fluoride by means of a nucleophilic substitution with 'hard' fluoride anion, the reaction will proceed most readily if the 'soft' Hg^{2+} cation is used together with the iodide anion as a 'soft' leaving group:

$$2 R . I + HgF_2 \rightarrow 2 R . F + HgI_2$$

Similarly in the preparation of iodides, the 'soft' iodide anion is preferably used as the nucleophile in association with a 'hard' potassium or sodium cation, and a 'hard' leaving group such as fluoride or chloride. For the same reason only 'hard' metal ions such as Zr^{4+}, Th^{4+} and Ti^{4+} catalyse the hydrolysis of alkyl fluorides.

Conversely, nucleophilic substitutions may be markedly retarded by metal ions that have a much higher affinity for the nucleophile than for the leaving group. Cyclisation of the bromide (I) to the aziridine (II), for example, is strongly inhibited by Cu^{2+} ions which bind the nucleophilic amino group and the imidazole nitrogen in the form of the chelate (III).

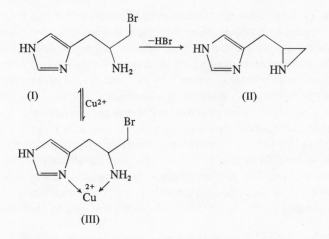

An additional way in which complex formation can facilitate a nucleophilic substitution at a saturated carbon is illustrated by systems in which, as a result of complex formation, the nucleophile is able to attack intramolecularly the saturated carbon at which the substitution occurs. Examples to illustrate this type of substitution and also the principles outlined above are discussed in the remainder of this section. The examples are largely of metal-ion catalysed nucleophilic substitutions of alkyl halides, and cyclic and acyclic ethers, because these types of compounds are so predominant in nucleophilic substitutions that have synthetic importance. The principles discussed, however, may be readily applied to other types of organic compounds.

Before considering specific examples, it should be noted that certain organometallic systems undergo S_N1 substitutions in which the intermediate carbonium ions are highly stabilised on account of the positive charge being delocalised by the metal (cf. p. 45). Because of this stability the substitutions proceed extremely rapidly, and treatment of the methyl ether $\eta\text{-}C_5H_5Fe(CO)_2$. CH_2OMe with hydrogen chloride in pentane, for example, causes immediate formation of the corresponding chloride $\eta\text{-}C_5H_5Fe(CO)_2$. CH_2Cl *via* the stable carbonium ion (IV). The high stability of the ion (V) similarly accounts for the rapidity of nucleophilic substitutions that take place at the α-positions of ferrocenes (see review by Cais (1966)).

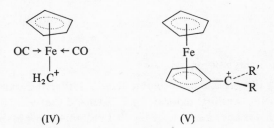

(IV)   (V)

A useful feature of these metal-stabilised carbonium ions is that they can often be generated by reactions other than those involving the departure of conventional leaving groups from saturated carbon, e.g. by hydride abstraction or protonation of an unsaturated ligand. In some cases the ions are sufficiently stable for them to be isolated and stored in the form of salts which may subsequently be subjected to attack by nucleophiles.

The complexed cyclohexadienyl cation (VI) has been extensively studied in this manner, and has been found to react with a very wide range of nucleophiles including D^-, RO^-, NC^-, enamines, conjugate bases of β-dicarbonyl compounds, and reactive aromatic systems. The iron tricarbonyl residue can easily be removed from the product of nucleophilic attack (VII) by mild oxidation (see p. 92), thus allowing the synthesis of many compounds difficult to obtain by other routes (see discussion by Birch and Jenkins (1976)).

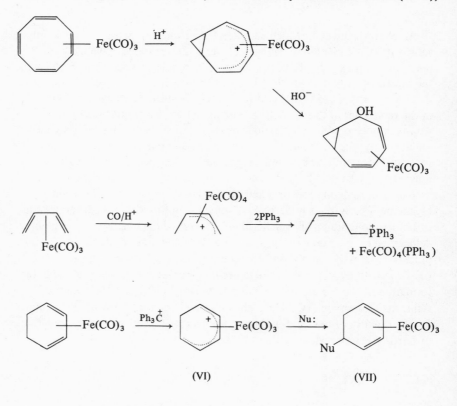

3.2.   Friedel-Crafts alkylations

The observation that a Lewis acid can promote the alkylation of a benzenoid system by an alkyl halide was first reported by Friedel and Crafts in 1877, when they showed that if benzene is treated with 1-chloropentane in the presence of aluminium trichloride, 1-phenylpentane is one of the hydrocarbons formed. Since then the use of Lewis acids and also proton donors to effect substitutions into aromatic substrates with many types of organic compounds including alkenes, ethers and alcohols has been extensively investigated. These Friedel–Crafts and related reactions have been discussed in great detail in a series of books edited by Olah (1963).

It is generally agreed that all the Friedel–Crafts reactions involve attack by an electrophile on a nucleophilic aromatic system, and they are accordingly classified as electrophilic substitutions. For convenience, however, Friedel–Crafts alkylations with alkyl halides are considered here as an example of nucleophilic substitution at a saturated carbon, with the aromatic substrate as the nucleophile.

$$Ar . H + R . X \rightarrow Ar . R + H . X \qquad\qquad [3.3]$$

The alkylation of an aromatic compound by an alkyl halide (equation 3.3) proceeds in the presence of a variety of inorganic compounds, e.g. zinc chloride, tin(IV) bromide, boron trifluoride and aluminium trichloride, of which the last is undoubtedly the one most frequently used in preparative work. Gallium trihalides are favoured for kinetic studies because of their comparatively high solubilities in organic solvents, and the lack of side reactions in the alkylation. All of these inorganic compounds function as Lewis acids, and with alkyl halides they form complexes in which the halogen is co-ordinated with the metal ion of the Lewis acid. Some of these complexes such as $Me_3C . F \rightarrow SbF_5$ are stable solids, but others, e.g. $Et . Cl \rightarrow GaCl_3$, are only capable of existence in solution, and their formation is deduced from physical evidence such as vapour pressure measurements and spectroscopic data. Several metal halides that are used as Friedel–Crafts catalysts are dimeric in the solid state with the monomer units held together by halogen bridges, but in the presence of an alkyl halide the dimeric structure is broken down, and the monomer functions as the active catalyst:

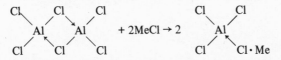

The co-ordination of the halogen of the alkyl halide (R . X) with the Lewis acid (M) increases the polarisation of the carbon–halogen bond, and the alkyl residue may acquire substantial carbonium-ion character (VIII*b*).

$$(R-X \rightarrow M \leftrightarrow R^+ \ X^- \rightarrow M) \rightleftharpoons (R)^+ \ (X-M)^- \rightleftharpoons (R)^+ + (X-M)^-$$

$$\text{(VIII}a) \qquad\qquad \text{(VIII}b) \qquad\qquad \text{(IX)} \qquad\qquad \text{(X)}$$

In solution the complex may ionise to give either a tight ion-pair (IX) or a solvent-separated ion-pair (X), with the extent of the ionisation being dependent upon the structure of the alkyl halide, the solvent and the Lewis acid. With alkyl fluorides and the very strong Lewis acid, antimony pentafluoride, complete ionisation occurs, especially if the pentafluoride (m.p. 6 °C) is also used as the solvent or co-solvent in conjunction with liquid sulphur dioxide. This approach was first used in 1961 to generate and observe spectroscopically the carbonium ion Me_3C^+, and has since been used to generate many other types of carbonium ions (see review by Olah (1973)), including the controversial norbornyl cation. With the slightly weaker Lewis acid, $AlCl_3$, ionisation still occurs even with primary halides, which in the presence of this acid rearrange by the 1,2-hydride and alkyl shifts so characteristic of carbonium-ion intermediates. n-Propyl chloride, for example, rearranges to isopropyl chloride.

$$Me \cdot CH_2 \cdot CH_2 \cdot Cl \xrightleftharpoons{AlCl_3} Me \cdot CH_2 \cdot CH_2 \cdot Cl \rightarrow AlCl_3 \rightleftharpoons Me \cdot CH_2 \cdot \overset{+}{C}H_2 + (AlCl_4)^-$$

$$Me_2CH \cdot Cl \xrightleftharpoons[-AlCl_3]{} Me_2CH \cdot Cl \rightarrow AlCl_3 \rightleftharpoons Me \cdot \overset{+}{C}H \cdot Me + (AlCl_4)^-$$

This isomerisation definitely involves the carbonium ions shown, and does not involve the reversible elimination and addition of hydrogen chloride, for no deuterium is incorporated if the isomerisation is carried out in the presence of deuterium chloride.

In accord with the formation of carbonium ions, optically active secondary and tertiary halides are readily racemised by Friedel–Crafts catalysts.

When complexed with a Lewis acid an alkyl halide is more susceptible to nucleophilic attack, particularly if the complex is appreciably ionised, and in the presence of an aromatic substrate nucleophilic substitution occurs to give a resonance-stabilised carbocation from which the alkylated product is formed by loss of a proton.

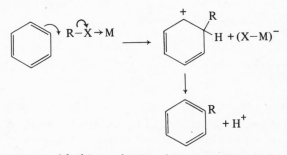

In agreement with this mechanism, kinetic studies on the alkylation of benzene and toluene by p-nitrobenzyl chloride and 3,4-dichlorobenzyl chloride in the presence of aluminium trichloride or tribromide give the rate equation:

$$\text{Rate} = k \, [ArH] \, [RCl] \, [AlX_3]$$

This does not reveal whether nucleophilic attack by the aromatic substrate is on the complex (VIII*a* ↔ VIII*b*) or on the ion-pair (IX), but the fact that the rate is first order in ArH does indicate that if the ion-pair is involved its formation by ionisation is not the rate-determining step of the reaction. This is also indicated by the rates of reaction being relatively insensitive to changes in the polarity of the solvent. Nucleophilic attack by the aromatic substrate on the carbonium ion of the solvent-separated ion-pair (X) is ruled out by product analysis. For example, in the alkylation of monosubstituted benzenes with a series of alkyl halides RCl, RBr and RI, the ratio of *ortho*- to *para*-disubstituted product is dependent on the halogen, and this would not be so if only the solvated carbonium ion were involved in the alkylation step. Nucleophilic attack on the un-ionised alkyl halide/Lewis acid complex is most likely to occur when ionisation would lead to unstable carbonium ions, e.g. Me^+ and Et^+. Consequently, although aluminium trichloride slowly converts 2-^{14}C-ethyl chloride into the 1-^{14}C compound *via* the carbonium ion ($^+CH_2 . CH_3$), the aluminium trichloride ethylation of benzene with the 2-^{14}C chloride gives no 1-^{14}C labelled ethyl benzene. Similarly, in studies of the gallium trichloride alkylation of a series of monosubstituted benzenes, the *ortho* and *para* partial rate factors vary in the order methylation > ethylation < t-butylation, and this is in accord with the first two of these processes proceeding by nucleophilic attack on the complex rather than on the carbonium ion as in the t-butylation.

In Friedel–Crafts alkylations where the rate of nucleophilic attack by the aromatic substrate is substantially lower than the rate of ionisation of the alkyl halide/Lewis acid complex, the carbonium ion may rearrange by hydride and alkyl shifts to an appreciable extent before reacting with the nucleophile. As a result a mixture of products is formed, and in extreme cases none of the expected product is observed. In the alkylation of benzene with 1-chloro-2,2-dimethylpropane, 2-chloro-2-methylbutane and 2-chloro-3-methylbutane, catalysed by aluminium chloride, an identical mixture of alkylated benzenes is obtained in each case, because of the rapidity with which the carbonium ions equilibrate.

$$Me_3C \cdot \overset{+}{C}H_2 \longrightarrow Me_2\overset{+}{C} \cdot CH_2Me \; \rightleftharpoons \; Me_2CH \cdot \overset{+}{C}HMe$$

$$\downarrow PhH \qquad\qquad\qquad\qquad \downarrow PhH$$

$$Ph \cdot CMe_2 \cdot CH_2Me \qquad Ph \cdot CHMe \cdot CHMe_2$$

With these reactions in which the intermediate carbonium ion rearranges, the composition of the product is determined by the relative rates at which the complex and the ion-pairs undergo nucleophilic attack, the rate at which ionisation occurs, and the rates at which the carbonium ions are interconverted. The rates of nucleophilic attack are partly dependent on the nucleophilicity of the aromatic substrate, and consequently the order of reactivity benzene < toluene < *m*-xylene < mesitylene is observed. Phenols and their ethers and aromatic amines are not nearly as reactive in Friedel–Crafts alkylations as they are in other reactions that involve electrophilic attack on the aromatic ring; the reactivity of *N,N*-dimethylaniline, for example, is comparable with that of nitrobenzene. This is because these types of compounds complex with the Lewis acid, as in (XI), and the lone pair of electrons which is normally

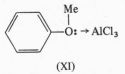

(XI)

responsible for activation of the aromatic ring towards electrophilic attack is directly involved in bond formation with the Lewis acid. An additional result of this complex formation is that the *ortho*-position of the benzene ring is sterically hindered, and hence the *ortho/para* ratio is decreased.

As most of the Friedel–Crafts catalysts which are commonly used are compounds of class '*a*' metals, their affinities for halide anions decrease in the order fluoride > chloride > bromide > iodide. Consequently this is also the order of the reactivities of the alkyl halides in the Friedel–Crafts alkylation of an aromatic substrate.

3.3. Silver-ion catalysis

In Friedel–Crafts alkylations with alkyl halides the aromatic substrates are fairly weak nucleophiles, and therefore comparatively strong Lewis acids have to be used to cause the extensive polarisation of the carbon–halogen bond that is necessary for the alkyl halide to undergo nucleophilic attack by the substrate. In nucleophilic substitutions of alkyl halides with much stronger nucleophiles, such as hydroxyl, cyanide and nitrite anions, fairly weak Lewis acids may be used to promote the reaction. The silver cation, in the form of a salt with the nucleophile as the gegenion, is often used for this purpose (equation 3.4). With some nucleophiles, e.g. dimethyl sulphoxide which is used for preparing aldehydes from primary halides (equation 3.5), the silver ion is most conveniently used as a salt with a very weak nucleophilic anion such as perchlorate as the gegenion.

$$R \cdot X + Ag \cdot Nu \rightarrow R \cdot Nu + AgX \qquad [3.4]$$

$$Me_2SO + R \cdot CH_2X + Ag^+ \rightarrow Me_2\overset{+}{S} \cdot O \cdot CH_2R + AgX$$
$$\downarrow$$
$$Me_2S + R \cdot CHO + H^+ \qquad [3.5]$$

It has often been stated or implied that in the presence of the silver cation, alkyl halides – even primary ones – react by an S_N1 mechanism. However, it is clear from kinetic and stereochemical studies that most silver-ion catalysed nucleophilic substitutions (and related rearrangements) are concerted, although bond breaking is often substantially more advanced than bond formation in the transition state.

A typical example of silver-ion catalysis is provided by the formation of N-alkylpyridinium salts in the reaction of primary and secondary bromides and chlorides with pyridine (used as the solvent). In the absence of silver ions the mechanism is S_N2, and a small amount of alkene formation is observed with the secondary halides but not with the primary ones. The salt formation is catalysed by silver nitrate but the mechanism of the catalysed reactions is definitely not S_N1, for the rates of reactions are not increased by the addition of small amounts of water, ethanol, or phenol; the primary halides react faster than the secondary, and no alkenes are formed from the former; and the extent of alkene formation that occurs with the secondary halides is the same as in the uncatalysed reaction. Evidently the silver ions catalyse the reaction by weakening the carbon–halogen bond, but not to an extent that causes the alkyl halides to ionise completely before nucleophilic attack occurs.

In the catalysed reactions described above, the 'push' of the nucleophilic pyridine is much greater than the 'pull' of the electrophilic silver cation, but when weaker nucleophiles are involved the two effects become comparable, and in some cases the 'pull' of the silver is greater than the 'push' of the nucleophile. When this arises the substitution is still concerted, but the alkyl residue acquires some carbonium-ion character in the transition state (XII) because bond breaking is ahead of bond formation, i.e. the alkyl halide ionises but only under the influence of the attacking nucleophile.

$$\text{Nu:} \quad \text{R--X} \quad \text{Ag}^+ \rightarrow \overset{\quad\quad \delta+ \quad \delta- \quad \delta+}{\text{Nu} \cdots \text{R} \cdots \text{X--Ag}} \qquad \text{(XII)}$$
$$\downarrow$$
$$\text{Nu--R} \quad \text{X--Ag}$$

As expected from a concerted reaction having S_N1 character, the extent of alkene formation which accompanies silver-ion catalysed nucleophilic substitutions is dependent on the halide used, in contrast to what would be observed if the reaction proceeded by a two step S_N1 mechanism in which the fate of the carbonium ion is independent of the halide anion. An example of this effect is demonstrated by the reaction of silver nitrate with 2-octyl halides in acetonitrile. The kinetics of the reaction show that the transition state is composed of nitrate anion, the alkyl halide and silver nitrate, but the yields of alkenes observed with 2-octyl bromide and 2-octyl chloride are 7.9 and 3.3 per cent respectively.

A further consequence of silver-ion catalysed reactions being concerted is that the substitution occurs mainly with inversion. This has recently been utilised in a modification of the classical Königen–Knorr reaction for preparing β-glucosides from the corresponding α-D-glucopyranosyl bromide. A mixture of the bromide and the appropriate alcohol is treated with the silver salt of 4-hydroxyvaleric acid in ether at −5 °C; excellent yields of the glucoside are obtained after a reaction time of 10 minutes.

In some nucleophilic substitutions the nucleophile is ambident, i.e. it can attack the alkyl halide through two different atoms, and therefore afford two different products. The nitrite anion, for example, can attack through oxygen or nitrogen, to give alkyl nitrites ($R-O-N=O$) and nitroalkanes ($R \cdot NO_2$) respectively. It is found that if these nucleophiles are used in the form of a silver salt the proportion of the product in which the nucleophile has attacked through the atom of higher electronegativity, i.e. oxygen in the case of the nitrite anion, is higher than when the nucleophile is used in the form of an alkali metal or tetra-ammonium salt. It is generally believed that this is because use of a silver salt increases the $S_N 1$ character of the substitution as described above, and somehow this in turn encourages the nucleophile to attack through the atom of higher electronegativity. In support of this explanation it is observed that in the reaction of an alkyl halide with nitrite anion as nucleophile the ratio of alkyl nitrite to nitroalkane is higher in acetonitrile than in ether, and varies with the type of alkyl halide in the order tertiary > secondary > primary. One point which appears to have been completely overlooked, however, is that in the silver salts the anion is bonded to the metal through the atom of *lower* electronegativity, e.g. nitrogen in NO_2^- and carbon in CN^-, and that the experimental observations can be explained equally well by assuming that the nucleophile involved is the undissociated silver salt rather than the free anion.

$$Ag-C\equiv N\colon \; + \; R\cdot X \rightarrow Ag\cdot CN \longrightarrow \; Ag\cdot C\equiv \overset{+}{N}\cdot R \; + \; [X\cdot Ag\cdot CN]^-$$

$$\bar{C}\equiv\overset{+}{N}\cdot R \; + \; Ag\cdot CN + Ag\cdot X$$

In the reactions with silver nitrite a five-centred intramolecular reaction is possible.

When silver-ion catalysed nucleophilic substitutions of alkyl halides are carried out in solvents in which the resultant silver halide is insoluble, e.g. aqueous ethanol, there is kinetic evidence that shows that the substitution occurs mainly on the surface of the silver halide which is precipitated during the reaction, and on which silver cations are absorbed.

3.4. Cleavage of co-ordinated acyclic ethers

In the acid-catalysed cleavage of ethers, protonation of the ether oxygen polarises the flanking oxygen–carbon bonds and also results in the formation of a more stable leaving group. The conjugate acid therefore undergoes nucleophilic attack at the α-carbon atoms much more readily than does the unprotonated ether. The same effect arises when the ether is co-ordinated to a metal ion, and in synthetic work ethers are often cleaved by treatment with a Lewis acid.

$$R-O-R' + M \rightleftharpoons \underset{\underset{M}{\downarrow}}{R-O-R'} \xrightarrow{Nu:} R'-Nu + R-\bar{O} \rightarrow M \qquad [3.6]$$

Halides of boron and aluminium are commonly used, particularly in the regeneration of a phenolic hydroxyl group that has been protected by conversion into a methyl ether. In these cases the nucleophile involved in the second step of the reaction is the halide anion.

$$Ar-O-Me + AlCl_3 \rightleftharpoons \underset{\underset{AlCl_3}{\downarrow}}{Ar-O-Me} \xrightarrow{Cl^-} Me \cdot Cl + \underset{\underset{Ar \cdot OH}{\downarrow H_3O^+}}{Ar \cdot O \cdot AlCl_2}$$

The equilibrium position for formation of a metal complex of an ether (the first step of equation 3.6) varies greatly. The most stable complexes are those in which the ether is co-ordinated to a class 'a' metal, and a substantial number of complexes formed by the halides of aluminium, gallium, magnesium, tin(IV) and titanium(IV) have been obtained as stable, crystalline solids. This is also true of several complexes of the type $R' \cdot MgX, 2R_2O$ formed by Grignard reagents. Most ethers complex readily with the boron trihalides, whose acceptor strengths increase in the order fluoride < chloride < bromide (see p. 59.) Although the trifluoride is the weakest acceptor of the three, its complexes with simple ethers are sufficiently stable to be distilled without decomposition, but if the donor strength of the ether oxygen is reduced by electron withdrawal or by delocalisation of the oxygen lone pair of electrons by resonance, as in di-(1-chloroethyl)ether and anisole respectively, the complexes dissociate into their two components when heated.

Many of the complexes formed from ethers and Lewis acids are stable

with respect to dissociation into the two components but are thermally
unstable, and when heated or kept at room temperature they decompose
according to equation 3.6 and give the products of ether cleavage. This
decomposition proceeds more readily if the ether oxygen is co-ordinated to
the Lewis acid as one of the donors of a chelated ligand, and consequently
the cleavage of aromatic ethers is accelerated by the presence of a co-ordinating
substituent in the *ortho*-position. This effect has been utilised for achieving
selective dealkylation of aromatic di-ethers.

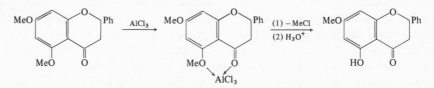

The ease with which an ether/metal halide complex decomposes is determined by the halide, and increases in the order fluoride < chloride <
bromide < iodide. This is because the decomposition proceeds by nucleophilic attack by the halide anion on the α-carbon atoms of the ether, and the
nucleophilicity of the halide anion is one of the factors that determines the
rate of this step. Except when a dissociative ionic mechanism is strongly
favoured, as in the cleavage of benzyl and benzhydryl ethers, nucleophilic
attack and departure of the leaving group are probably concerted processes
(with attendant inversion of configuration) in which the alkyl group acquires
some carbonium ion character. An intramolecular process of the S_Ni type
(equation 3.7) which would involve retention of configuration, also seems

[3.7]

feasible as a reaction mechanism, but since an optically active alkyl halide
produced would be racemised by the Lewis acid (see p. 128) at a faster rate
than the ether would be cleaved, it would appear that attempts to distinguish
between the two types of mechanism by comparing the stereochemistry of
an optically active ether with that of the alkyl halide obtained from it would
be uninformative.

3.5. Cleavage of co-ordinated acetals and ketals

Acetals and ketals are often used as protecting groups for aldehydes and
ketones respectively, because as ethers they are inert to nucleophilic reagents,
but under acidic conditions they are rapidly hydrolysed to the parent
carbonyl compound through a resonance-stabilised carbonium ion and the
corresponding hemi-acetal or hemi-ketal.

$$R \cdot CH(OEt)_2 \underset{H^+}{\rightleftharpoons} \underset{\underset{+}{\overset{|}{H \cdot OEt}}}{R \cdot CH - OEt} \underset{-EtOH}{\rightleftharpoons} [R - \overset{+}{C}H - OEt \leftrightarrow R - CH = \overset{+}{O}Et]$$

$$\downarrow H_2O$$

$$R \cdot CHO + EtOH \longleftarrow \underset{\overset{|}{OH}}{R \cdot CH - OEt}$$

The same carbonium ion is also formed – probably as a component of an ion-pair – when acetals (and ketals) are treated with strong Lewis acids, and in the presence of boron trifluoride an acetal may be added across the double bond of a vinyl ether to give a β-alkoxyacetal.

$$R \cdot CH(OEt)_2 \underset{BF_3}{\rightleftharpoons} \underset{\overset{|}{EtO \to BF_3}}{R \cdot CH - OEt} \rightleftharpoons R - \overset{+}{C}H - OEt + [EtO \cdot BF_3]^-$$

$$\downarrow H_2C:CH \cdot OEt$$

$$R \cdot CH(OEt) \cdot CH_2 \cdot CH(OEt)_2 \underset{[EtO \cdot BF_3]^-}{\rightleftharpoons} R \cdot CH(OEt) \cdot CH_2 \cdot \overset{+}{C}H \cdot OEt$$
$$+ BF_3$$

This type of addition has been utilised for chain extensions in the carotenoid series; for example, in one of the commercial syntheses of β-carotene the C_{14}-aldehyde (XIII) is converted into the C_{16}-aldehyde (XIV).

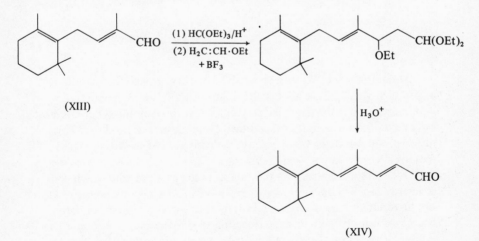

(XIII)

(1) $HC(OEt)_3/H^+$
(2) $H_2C:CH \cdot OEt$
$+ BF_3$

H_3O^+

(XIV)

The magnesium halide which is invariably present in an ether solution of a Grignard reagent also promotes the ionisation of the carbon–oxygen bonds of acetals, and the ring contraction that occurs in the reaction between 6-ethoxy-Δ^2-dihydropyran and Grignard reagents involves the carbonium ion formed from the complexed pyran.

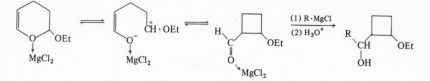

Because of the presence of the magnesium halide, acetylenic Grignard reagents react with acetals to give ethers (equation 3.8) although the corresponding sodium derivatives which are stronger nucleophiles do not. The analogous reaction with ethyl orthoformate is used in a preparation of aldehydes (equation 3.9), and the stereochemistry of this type of reaction has been deduced from a study of the reaction between Grignard reagents and 4,6-disubstituted-2-methoxy-1,3-dioxanes of known stereochemistry.

$$R \cdot C \equiv C \cdot MgBr + Et \cdot CH(OEt)_2 \xrightarrow{MgBr_2} R \cdot C \equiv C \cdot CHEt \cdot OEt \qquad [3.8]$$

$$R \cdot MgBr + HC(OEt)_3 \xrightarrow{MgBr_2} R \cdot CH(OEt)_2$$
$$\downarrow H_3O^+$$
$$R \cdot CHO \qquad [3.9]$$

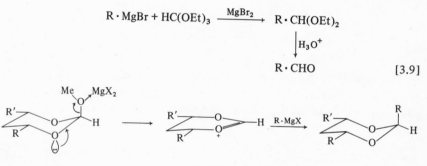

The conformers (XV) with an axial 2-methoxyl group react rapidly to afford high yields of the 2-substituted dioxan (XVII) with retention of configuration, but the conformers with an equatorial methoxyl group react only under forcing conditions, and then give mixtures of products. This difference in reactivity arises because the departure of the co-ordinated methoxyl group occurs by an antiparallel elimination which involves the axial lone pair of electrons on the adjacent oxygen atom, and which consequently requires the methoxyl group to be in an axial environment. Attack on the resultant carbonium ion (XVI) by the Grignard reagent similarly occurs exclusively from an axial direction to give the chair conformer (XVII).

In these reactions between Grignard reagents and ortho esters the slowest stage is the formation of the carbonium ion from the co-ordinated ester, e.g. (XV) → (XVI), and the overall rate of reaction can therefore be increased by increasing the concentration of co-ordinated ester. This can be accomplished by modifying the alkoxy group that is involved in co-ordination with the magnesium halide so that its affinity for magnesium is raised. Consequently,

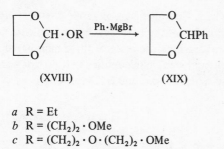

(XVIII)   (XIX)

a R = Et
b R = (CH$_2$)$_2$ · OMe
c R = (CH$_2$)$_2$ · O · (CH$_2$)$_2$ · OMe

the reactivity of the dioxolanes (XVIII; *a*—*c*) increases in the order *a* < *b* < *c*, and under the conditions where (XVIII*c*), which has a tridentate chelating alkoxy group, reacts with phenylmagnesium bromide to give an excellent yield of 2-phenyl-1,3-dioxolane (XIX), the ethoxy analogue (XVIII*a*) is largely recovered. The formation of the carbonium ion from the complexed ortho ester is also facilitated, of course, when the ester oxygen which is co-ordinated with the magnesium halide is that of a good leaving group. Consequently, the phenoxy analogue PhO . CH(OEt)$_2$ reacts with Grignard reagents much faster than does ethyl orthoformate, and is now preferred to the latter reagent in the preparation of aldehydes.

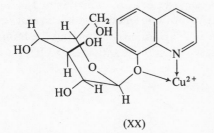

(XX)

Mechanistically related to these substitutions is the copper(II)-catalysed hydrolysis of 8-hydroxyquinoline-β-D-glucoside. The hydrolysis of acetals, ketals and simple glycosides is not subject to metal-ion catalysis because of the low affinity of the ether oxygens for metal ions relative to that of the water, which is necessarily present to cause hydrolysis. The additional binding site in the quinoline glucoside overcomes this adverse factor (see (XX)), and the value of the observed first-order rate constant for the hydrolysis at 70 °C is almost the same with [copper(II) ions] = 1.25 × 10^{-3} M as with [HCl] = 4.18 M.

3.6.   Cleavage of co-ordinated oxiranes

The complexes formed between Lewis acids and simple five- and six-membered ring ethers are more stable and less reactive towards nucleophiles than those of analogous straight-chain ethers. Complexes of oxiranes, however, are far more reactive because co-ordination of the heterocyclic oxygen with the Lewis acid increases the ease with which the reactive three-membered ring is opened by nucleophilic attack, and weak nucleophiles that do not normally react with oxiranes may do so when the oxirane is complexed.

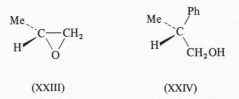

(XXI)   (XXII)

The proportions of the two possible products (XXI) and (XXII) formed in these promoted ring-opening reactions vary with the Lewis acid, the oxirane, and the reaction conditions. Factors that are conducive to a high degree of charge separation along one of the carbon–oxygen bonds, either in the starting complex or in the transition state, lead to attack by the nucleophile on the carbon which can form the more stable carbonium ion, but factors that do not favour charge separation are usually associated with nucleophilic attack on the less hindered carbon. The almost exclusive formation of 1-iodopropan-2-ol when propylene oxide is treated with magnesium iodide in ether is indicative of a concerted, non-ionic $S_N 2$ type of mechanism, but the formation of (R)-2-phenylpropan-1-ol (XXIV) when the enantiomer (XXIII) is used in a

(XXIII)   (XXIV)

Friedel-Crafts reaction with benzene and aluminium trichloride indicates a mechanism which is also concerted, but in which bond breaking has progressed substantially further than bond formation in the transition state. The latter mechanism is directly analogous to the $A2$ mechanism (equation 3.10) which is now favoured over the $A1$ mechanism (equation 3.11) for the acid-catalysed hydrolysis of oxiranes.

With co-ordinated oxiranes in which the heterocyclic oxygen forms part of a chelating system, nucleophilic attack may occur with almost complete regiospecificity. Thus in the presence of Cu^{2+}, methanol reacts with 2-pyridyloxirane to give exclusively the β-methoxy derivative, in contrast to the reaction with sodium methoxide in methanol which gives a 1 : 2 ratio of the α- and β-derivatives.

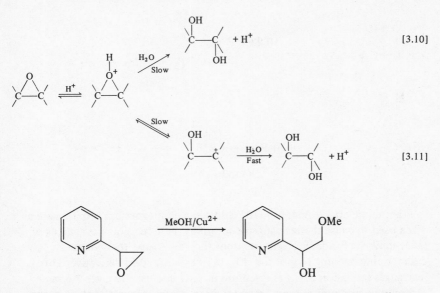

[3.10]

[3.11]

This specificity arises because the five-membered ring chelate initially produced by nucleophilic attack at the β-position is more stable than the corresponding six-membered ring chelate produced by α-attack.

In the absence of a nucleophile an oxirane complex may also be ring-opened by means of a 1,2-shift to produce an aldehyde or ketone. On account of the low nucleophilicity of the fluoride anion, boron trifluoride is often the Lewis acid used to induce this rearrangement, although lithium perchlorate in benzene with tri-n-butylphosphine oxide as a solubilising agent has recently been shown to be highly efficient and very rapid in action.

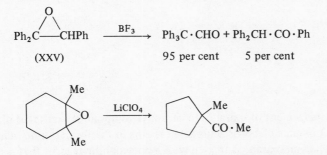

The rearrangement of an oxirane complex that bears four different substituents can theoretically give rise to four different carbonyl compounds, but in practice the reaction proceeds largely by the formation of the more stable of the two carbonium ions (XXVI) and (XXVII) and a 1,2-shift of one of the substituents. This shift may accompany the formation of the carbonium ion in a concerted reaction, or follow it as a discrete step as shown in figure 3.1.

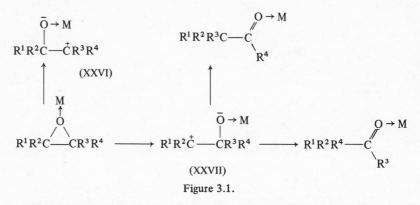

Figure 3.1.

Aluminium trichloride, iron(III) chloride, and magnesium bromide are also often used to cause the rearrangement, but as these reagents are sources of nucleophilic chloride or bromide anions they also cause the oxirane to undergo ring-opening to give the chloro- or bromohydrin derivative. This rearranges (equation 3.12) as discussed in §3.7 and usually gives the same carbonyl compounds as are formed by direct rearrangement of the oxirane but in different proportions. For this reason, the percentage composition of the mixture of carbonyl compounds obtained by treatment of an oxirane with a Lewis acid depends on which Lewis acid is used. The ratio of aldehyde to ketone formed from the oxirane (XXV), for example, changes from 19 : 1 to about 3 : 1 when aluminium trichloride is used in place of boron trifluoride.

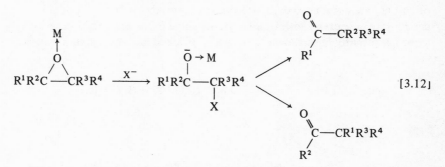

[3.12]

The rearrangement of oxiranes to carbonyl compounds that takes place under the influence of Lewis acids often results in the formation of 'abnormal' products when an oxirane is treated with a nucleophilic reagent that is also a Lewis acid. One very instructive example of this is in the reduction of the oxirane (XXVIII). With lithium aluminium hydride the expected alcohol (XXIX) is formed by hydride attack on the less hindered carbon atom, but when a 3 : 1 mixture of lithium aluminium hydride and aluminium trichloride is used the attack by hydride anion is on an aluminium complex of the oxirane. This complex presumably has some carbonium-ion character, for attack takes place at the more substituted carbon to give the secondary alcohol (**XXX**). In contrast to both these situations, the use of a 1 : 4 mixture of lithium aluminium hydride and aluminium trichloride causes the oxirane to isomerise to the aldehyde (**XXXI**) at a faster rate than it is reduced and the reduction therefore ultimately affords the primary alcohol (**XXXII**).

$$Me_2CH \cdot CH(OH) \cdot CMe_3 \xleftarrow[\substack{+\\AlCl_3}]{3\ LiAlH_4} Me_2C\overset{O}{\overbrace{\quad}}CH \cdot CMe_3 \xrightarrow{LiAlH_4} Me_2C(OH) \cdot CH_2 \cdot CMe_3$$

$$(XXX) \qquad\qquad\qquad (XXVIII) \qquad\qquad (XXIX)$$

$$\downarrow \substack{LiAlH_4\\+\\4\ AlCl_3}$$

$$Me_3C \cdot CMe_2 \cdot CH_2OH \longleftarrow Me_3C \cdot CMe_2 \cdot CHO$$

$$(XXXII) \qquad\qquad\qquad (XXXI)$$

Treatment of oxiranes with Grignard reagents in ether often affords 'abnormal' alcohols, which arise from the Grignard reagents reacting with the carbonyl compounds formed from the oxirane under the influence of the magnesium halide which is invariably present in the ether solution of the Grignard reagent. This difficulty can usually be overcome by use of the appropriate dialkyl or diarylmagnesium compound.

3.7. Conversion of halohydrins into carbonyl compounds

In the preceding section it was pointed out that one of the pathways for the conversion of oxiranes into carbonyl compounds, catalysed by Lewis acids, is initiated by the co-ordinated oxirane undergoing nucleophilic attack by a halide anion to give a halohydrin derivative which then rearranges.

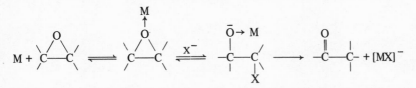

A number of metal halides bring about this conversion, but lithium bromide in benzene with tri-n-butylphosphine oxide or hexamethylphosphoramide as a solubilising agent is particularly effective; high yields of carbonyl compounds may be obtained after short reaction times.

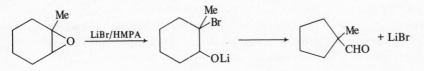

The second step of the reaction, i.e. the rearrangement of the halohydrin derivative, may be carried out independently of the first step by heating the halomagnesium derivative (XXXIII; M = MgX) prepared from the parent halohydrin by the action of a simple Grignard reagent, and most of the studies on the stereochemistry of the rearrangement step have used this type of derivative. It is reasonable to assume that in non-co-ordinating solvents at least, these halomagnesium derivatives exist as chelates in which the halogen is coordinated with the magnesium (as shown in XXXIII), and that this coordination assists the departure of the halogen and the attendant migration of the adjacent substituent.

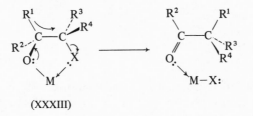

(XXXIII)

For the halohydrin derivative (XXXIII) to rearrange it is necessary for the halogen and the migrating group to have an anti-periplanar relationship, and consequently although the bromomagnesium derivative (XXXIV) of *trans*-2-chloro-1-methylindanol steadily rearranges when heated in benzene to give the expected 2-methylindanone (XXXV), the stereoisomer in which the methyl group and the halogen are *cis* with respect to each other, and which cannot adopt the required stereochemistry, gives mainly polymeric material and only a trace of the ketone.

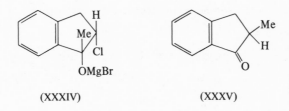

(XXXIV)   (XXXV)

The high stereospecificity of the rearrangements of halohydrin derivatives has been utilised in a convenient method for stereospecifically introducing an angular methyl group into the decalone system. In the addition of methyl-lithium to *trans*-9-chlorodecalone, the organolithium approaches the molecule from the side opposite to the halogen for steric reasons, and hence the methyl group in the resultant chlorohydrin (XXXVI) is *trans* with respect to the chlorine. The thermal rearrangement of the bromomagnesium derivative of this chlorohydrin therefore affords *cis*-9-methyldecalone (XXXVII).

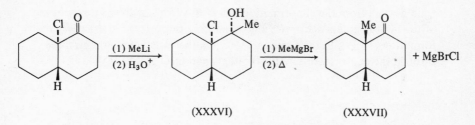

(XXXVI)   (XXXVII)

In the rearrangement of the derivative (XXXIII) of an acyclic halohydrin in which the substituents R^1 and R^2 are not identical, there are two conformations in which either R^1 or R^2 is antiperiplanar with respect to the halogen (as illustrated by the Newman projections (XXXIX*a*) and (XXXIX*b*), for example), and hence the rearrangement can theoretically give two different carbonyl compounds. In practice one of these predominates. Thus, both diastereomers of (XXXVIII) largely afford butan-2-one because a hydride anion undergoes a 1,2-shift more readily than does a methyl group.

$$Me \cdot CHBr \cdot CH(OMgBr) \cdot Me$$

(XXXVIII)

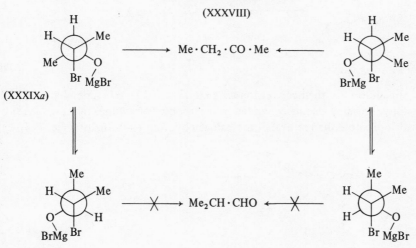

(XXXIX*a*)

(XXXIX*b*)

Exceptions arise, however, in systems where substantial steric compression occurs in the transition state of the rearrangement. This is often the case with halomagnesium derivatives, for the oxymagnesium halide group behaves conformationally as a very bulky group – presumably owing to solvation of the magnesium ion – and its presence can have a marked stereochemical effect on the rearrangement. In the rearrangement of the diastereomers of (XL; Ar = p-C_6H_4Cl, M = Mg Br), for example, the steric interaction between the oxymagnesium bromide group and the 2-phenyl group in the transition state derived from the conformation (XLI*a*) is sufficient to ensure that one of the diastereomers rearranges almost entirely through the alternative conformation (XLI*b*). For the same reason the other diastereomer rearranges through the conformation (XLII*a*) rather than through (XLII*b*).

Ph · CHBr · C(OM)ArPh

(XL)

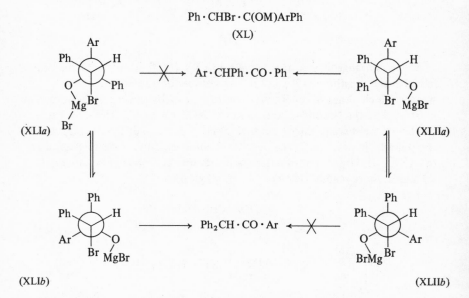

(XLI*a*) (XLII*a*)

(XLI*b*) (XLII*b*)

In addition to their conversion *via* a metal alkoxide as described above, halohydrins may also be converted into carbonyl compounds by Lewis acids which promote the removal of the halogen by prior co-ordination (see (XLIII)).

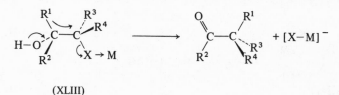

(XLIII)

Boron trifluoride and silver cations are commonly used for this purpose. As in simple nucleophilic substitutions the latter reagent promotes a non-ionic, concerted mechanism, but the former favours the formation of an intermediate carbonium ion and therefore tends to give mixtures of products. Thus although treatment of the diastereometric chlorohydrins (XLIV) and (XLV) with silver oxide in n-hexane affords only the ketone (XLVI) and the aldehyde (XLVII) respectively, treatment with boron trifluoride gives a mixture of both carbonyl compounds from each of chlorohydrins. It should be noted that in the rearrangements promoted by silver oxide, an antiperiplanar relationship between the departing halogen and the migrating group is rigorously maintained.

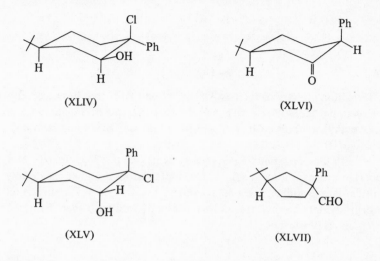

(XLIV)

(XLVI)

(XLV)

(XLVII)

It should also be noted that the silver-catalysed rearrangement of the diastereomers of the bromohydrin (XL; Ar = p-C_6H_4Cl, M = H) contrasts with the thermal rearrangement of their halomagnesium derivatives. Thus, while one of these derivatives rearranges to $Ph_2CH.CO.Ar$ through the conformation (XLIb), the parent bromohydrin rearranges to the isomeric ketone $Ar.CHPh.CO.Ph$ through the conformation corresponding to (XLIa). Similarly, while the other diastereomeric halomagnesium derivative gives $Ar.CHPh.CO.Ph$ through (XLIIa), the parent bromohydrin gives $Ph_2CH.CO.Ar$ through the conformation corresponding to (XLIIb). Evidently, in both cases the replacement of the bulky magnesium bromide group by hydrogen reverses the relative stabilities of the two transition states that result from the two possible conformations.

3.8. Cleavage of co-ordinated thioethers

Thioethers are cleaved in a manner identical to that of their oxygen analogues (e.g. equation 3.13), but their rates of cleavage by compounds of class 'a' metals are lower than those of the corresponding ethers because of the lower affinity of class 'a' metals for sulphur than for oxygen. Anisole is completely cleaved when heated at 200 °C for 1 hour with magnesium iodide or at 60 °C for 18 hours with aluminium tribromide in chlorobenzene, but under both sets of conditions phenyl methyl sulphide is recovered unchanged.

$$Ph.S.CH_2Ph + AlBr_3 \rightarrow Ph.S.AlBr_2 + Ph.CH_2Br \qquad [3.13]$$

In contrast, class 'b' metal ions readily promote the cleavage of thioethers by co-ordinating with the sulphur atom and making the α-carbon atoms more susceptible to nucleophilic attack.

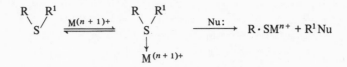

Transition metal ions such as Au^{3+}, Pt^{2+} and Pd^{2+} that have a d^8 configuration seem particularly effective, and the first example of S-demethylation (of dimethyl sulphide), which was reported in 1883 by Blomstrand, utilised platinum(II).

As with the corresponding oxygen systems (see p. 134), cleavage of a thioether proceeds most readily if the sulphur atom is one of the donors of a multidentate ligand, and several examples of very rapid S-demethylation of bi- or tri-dentate methylthio compounds by d^8 ions have been described, of which one is illustrated.

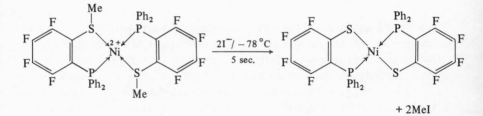

+ 2MeI

Co-ordination of a thioether with mercury(II), silver(I) or copper(I), all of which have d^{10} ions, also strongly promotes nucleophilic substitution at the α-carbons. Thioacetals and thioketals are conveniently converted into their oxygen analogues when heated with mercury(II) chloride in an alcohol; cadmium carbonate is often added to remove the hydrogen chloride formed:

$$R \cdot CH(SEt)_2 + 2MeOH \xrightarrow{HgCl_2} R \cdot CH(OMe)_2 + Hg(SEt)_2 + 2HCl$$

Although thioacetals and thioketals are stable to aqueous acids, rapid hydrolysis to the parent aldehyde or ketone occurs when they are heated with mercury(II) chloride or silver oxide in aqueous acetonitrile, and this has been used in the last step of a very useful synthesis of ketones from dithianes (see top of following page).

In the presence of copper(I) chloride, thioacetals and thioketals undergo nucleophilic attack by carbanions (e.g. equation 3.14), but in contrast to their oxygen analogues (see p. 136), orthothioesters, $HC(SR)_3$, do not react with Grignard reagents, in accord with the low affinity of the 'hard' magnesium(II) ion for 'soft' sulphur.

$$Ph \cdot CH(SEt)_2 + CH_2(CN)_2 \xrightarrow{Base/CuCl} Ph \cdot CH(SEt) \cdot CH(CN)_2 \qquad [3.14]$$

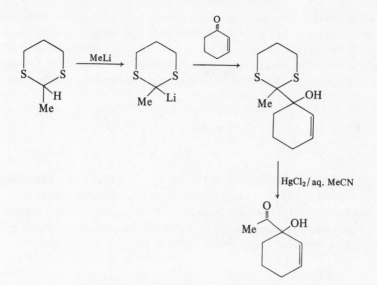

Because of the ability of sulphur to stabilise positively-charged carbon, it seems likely that these nucleophilic substitutions on co-ordinated thioacetals and thioketals have substantial S_N1 character. This is also true of the mercury(II)-promoted cleavage of benzylthio and t-butylthio ethers, a reaction used in the peptide field for the regeneration of the thiol function of S-protected cysteine residues.

$$
\begin{array}{ccc}
\underset{\displaystyle |}{CH_2 \cdot SR} & \xrightarrow[\text{in 80 per cent aq. AcOH}]{(CF_3 \cdot CO_2)_2 Hg} & \underset{\displaystyle |}{CH_2 \cdot S \cdot Hg \cdot O \cdot CO \cdot CF_3} \\
\text{\sim NH} \cdot CH \cdot CO \text{\sim} & & \text{\sim NH} \cdot CH \cdot CO \text{\sim}
\end{array}
$$

R = CMe$_3$
or CH$_2$Ar

$\downarrow$ H$_2$S

$$
\underset{\text{\sim NH} \cdot CH \cdot CO \text{\sim}}{\overset{\displaystyle |}{CH_2 \cdot SH}}
$$

Metal-promoted reactions of thioethers, thioacetals and thioketals have been discussed by Satchell (1977).

3.9. Intramolecular nucleophilic substitutions

In all the nucleophilic substitutions that have been discussed so far in this chapter, substitution is accelerated because the leaving group of the substrate is complexed with a metal ion. Complex formation may also cause a nucleophilic substitution at a saturated carbon to be accelerated if in the resultant complex the nucleophile can attack the saturated carbon intramolecularly. In most cases where this arises the nucleophile is one of the ligands attached to the metal, and the reaction may be represented by the general equation 3.15 where (L) is a donor group in the compound undergoing substitution. Although this type of intramolecular reaction usually involves a five- or six-membered ring transition state, four-membered ring transition states are quite common.

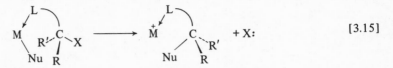

[3.15]

Reduction of a chlorohydrin $(Cl . (CH_2)_{n+1} . CH_2OH)$ or the corresponding acid $(Cl . (CH_2)_{n+1} . CO_2H)$ with lithium aluminium hydride to give the alcohol $(Me . (CH_2)_n . CH_2OH)$ proceeds very readily if $n = 0$ or 1 because of the participation of complexes of type (XLVIII), but when $n = 3$ or 4 the reduction is very slow and is almost certainly intermolecular.

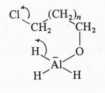

(XLVIII)

Nucleophiles such as secondary amines, and hydroxyl or bromide anions react with *trans*-4-hydroxy-1,2-epoxycyclohexane to give the 4-substituted *trans*-1,3-diol by axial attack at C_1 of the stable conformation (XLIX). With lithium aluminium hydride, however, complex formation results in intra-molecular hydride transfer to C_2 to give the *trans*-1,4-diol, even though the intermediate complex involves the unfavourable conformation shown in (L).

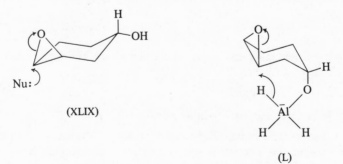

(XLIX)   (L)

An even more striking example of how complex formation can alter the position of nucleophilic attack is provided by reactions of the hexahydroisochromone (LI), obtained when carvenone reacts with benzaldehyde in the presence of hydrogen chloride. Treatment of the isochromone with sodium methoxide or sodium hydroxide causes dehydrochlorination and the formation of the expected αβ-unsaturated ketone, but treatment with zinc methoxide or magnesium methoxide affords the mixed acetal (LIII) by a type of reaction which is rarely met (although the analogous S_N2' reaction is often encountered with allylic compounds), but which occurs in this particular system because in the intermediate complex (LII) the nucleophilic methoxyl group is correctly situated for attack on the carbon atom that is in the γ-position with respect to the leaving group.

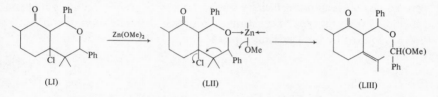

(LI)   (LII)   (LIII)

It is evident that if a nucleophilic substitution is to proceed through a complex as represented by equation 3.15, the carbon atom that bears the leaving group must be near a donor group (L) that can co-ordinate with the metal ion. The ease with which a complex is formed – and consequently the degree of intramolecularity of the substitution process – will depend on the co-ordinating strength of this donor group, and will be greater if the group is part of a chelating system. In agreement with this, the reactivity of the bromides (LIV; *a–d*) towards nucleophilic reagents of the type Nu . M (M = Li or MgX) increases in the order $a < b < c < d$ which is in accord with their ability to enter into complex formation. The bromide (LIV*d*), in marked contrast to most saturated bromides, even reacts fairly readily with phenylethynylmagnesium bromide to give the disubstituted alkyne Me(O . CH$_2$. CH$_2$)$_3$. C : C . Ph, presumably *via* the chelate (LV).

L · (CH$_2$)$_2$Br

a   L = n-Pr
b   L = MeO
c   L = MeO · (CH$_2$)$_2$ · O
d   L = MeO · [(CH$_2$)$_2$ · O]$_2$

(LIV)   (LV)

There are a number of nucleophilic substitutions that have been suggested to proceed by intramolecular routes in which the functional group responsible for complex formation is actually the leaving group. Although these mechanisms appear attractive, with the majority there is little practical evidence to substantiate them.

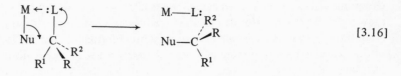

$$[3.16]$$

Support for the existence of the mechanism shown in equation 3.16 however, has been obtained from kinetic studies of the reaction between various reactive halides and the nickel(II) complexes of chelating thiols. For example, the data obtained from benzyl bromide and chelates of type (LVI) were consistent with the equilibrium formation of (LVII) followed by rate-determining nucleophilic attack by sulphur on carbon, but it has not yet been established that this type of substitution proceeds with retention of configuration as required by the mechanism [3.16].

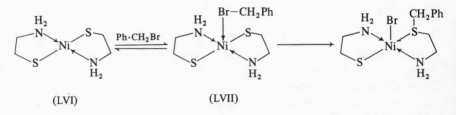

(LVI)   (LVII)

This last point is particularly relevant to the 'co-ordination-propagation' mechanism suggested for the polymerisation of oxiranes by catalysts such as iron(III) alkoxides (see Johnson, 1965, for a useful survey of this topic). This mechanism, the initial steps of which are illustrated in figure 3.2, requires the oxirane ring to open with retention of configuration, but in systems that have been amenable to stereochemical studies it has been found that the ring usually is opened with inversion.

$$\text{Me} \cdot \text{CH} \underset{O}{\overset{}{—\!—}} \text{CH}_2 + \text{Fe(OR)}_2\text{Cl} \longrightarrow \text{Me} \cdot \text{CH(OR)} - \text{CH}_2$$
$$\text{O} - \text{Fe(OR)Cl}$$

Figure 3.2.

Partial retention of configuration (~20 per cent) has, however, been observed in the ring-opening of (R)-styrene oxide by dialkyl magnesiums. The complex (LVIII) can evidently react intramolecularly to give (LIX), or react intermolecularly with a second molecule of dialkylmagnesium to give the enantiomer (LX).

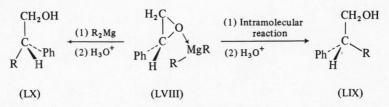

(LX)   (LVIII)   (LIX)

Complete retention of configuration has been observed in the reduction of *gem*-bromofluorocyclopropanes by lithium aluminium hydride in tetra-hydrofuran.

An intramolecular substitution with a six-membered ring transition state has been suggested on many occasions for the reaction between an alkyl halide and a metal enolate (LXI; L = 0) derived from an aldehyde or ketone.

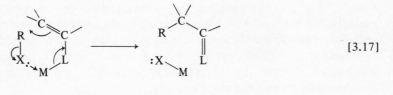

[3.17]

(LXI)

This reaction is extensively used for the alkylation of ketones, and in practice the alkyl halide is added to the metal enolate which is generated *in situ* either by treating the ketone with a base such as sodium hydride, triphenymethyl-lithium or lithium diethylamide, or indirectly from the reaction between an enol ester and two equivalents of methyl-lithium. The latter method has the advantage that a specific metal enolate, e.g. (LXII), can be generated from a ketone that can give two isomeric anions. (see p. 152.)

In an alternative procedure, the imine obtained from an aldehyde or ketone and t-butylamine is treated with a Grignard or organolithium reagent, and then the alkyl halide is added to the resultant magnesium derivative.

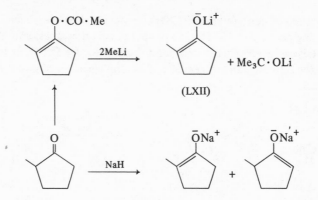

(LXII)

Hydrolysis of the Schiff base which is formed gives the alkylated product, as in the example shown.

$$Me_2CH \cdot CHO \xrightarrow{H_2N \cdot Bu^t} Me_2CH-CH=N \cdot Bu^t \xrightarrow{MeMgBr} Me_2C=CH-N \cdot Bu^t$$
$$\underset{MgBr}{|}$$

$$\downarrow EtBr$$

$$Me_2C(Et) \cdot CHO + H_2N \cdot Bu^t \xleftarrow{H_3O^+} Me_2C(Et)-CH=N \cdot Bu^t + MgBr_2$$

The possibility that this alternative procedure involves an intramolecular reaction of type [3.17] is supported by various stereochemical studies. For example, alkylation of the lithium derivative of the imine formed from cyclohexanone and the chiral amine (LXIII) affords a Schiff base which on hydrolysis gives the 2-alkylcyclohexanone with the R-ketone in an enantiomeric excess of 80–90 per cent, and this result can be shown to be completely consistent with the process [3.17].

(LXIII)

It is interesting that although in the apparently related reaction between oxiranes and allylic Grignard reagents the formation of rearranged products is consistent with an intramolecular mechanism, e.g. [3.18], substitution at the saturated carbon of the oxirane has been found to occur with inversion rather than with retention of configuration, and the alternative intermolecular mechanism, e.g. [3.19], seems far more likely.

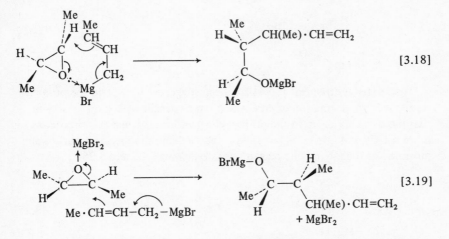

[3.18]

[3.19]

Reactions between allylic systems and organometallic or inorganic reagents often afford products that can be rationalised by an $S_N 2'$ type of mechanism [3.20], in which the metal not only holds the nucleophilic species close to the double bond but also facilitates departure of the leaving group.

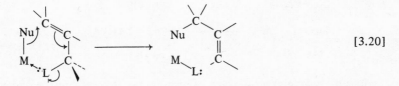

[3.20]

For example, the formation of vinyl ethers from Grignard reagents and the acetals of $\alpha\beta$-unsaturated aldehydes, and the reduction of certain allylic bromides by aluminium hydride, may be represented by equations 3.21 and 3.22 respectively. The rapidity with which allylic halides rearrange when treated with the corresponding halide of iron(III), zinc(II) or copper(II) is similarly explained by the intramolecular reaction 3.23.

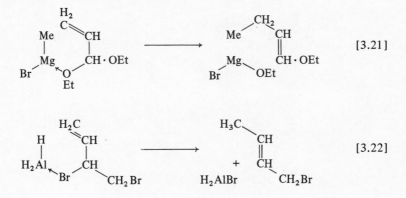

[3.21]

[3.22]

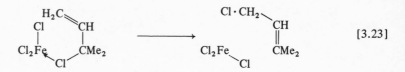

[3.23]

The related mechanism [3.24] has been suggested to explain the observation that in the formation of an allene by nucleophilic attack on an alkyne that has a leaving group in the propargylic position, the reaction proceeds more readily and gives a higher yield if the nucleophilic species is part of an inorganic or organometallic reagent, as in reactions 3.25 and 3.26.

$$R-C\equiv C-\overset{\overset{R^1}{|}}{\underset{\underset{}{|}}{C}}-R^2 \longrightarrow \overset{R}{\underset{Nu}{}}C=C=C\overset{R^1}{\underset{R^2}{}} \quad\quad [3.24]$$
$$Nu-M \leftarrow :L \quad\quad\quad + \quad M-L:$$

$$HC\equiv C-\overset{\overset{R^1}{|}}{\underset{\underset{}{|}}{C}}-R^2 \longrightarrow NC\cdot CH=C=CR^1R^2 \quad\quad [3.25]$$
$$NC-Cu \quad Br \quad\quad\quad + CuBr$$

$$MeC\equiv C-\overset{}{\underset{\underset{}{|}}{C}}Me_2 \longrightarrow Me_2C=C=CMe_2 \quad\quad [3.26]$$
$$Me-Li \quad Cl \quad\quad\quad + LiCl$$

In some examples of this type of reaction the stereochemistry is in agreement with the suggested mechanism. Thus although reduction of the tosylate (LXIV) with lithium aluminium hydride affords an optically active allene (LXV) whose configuration is consistent with the hydride anion approaching the triple bond from the side of the molecule opposite to the tosylate group, reduction of the parent alcohol gives the enantiomeric allene (LXVI) which suggests the involvement of an aluminium complex, as shown.

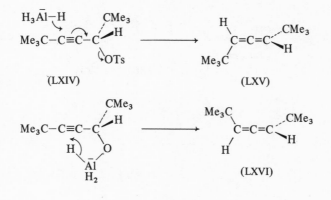

Related to the last reaction is the reduction by lithium aluminium hydride of mono(tetrahydropyranyloxy) derivatives of butyn-1,4-diols, which gives excellent yields of allenic alcohols.

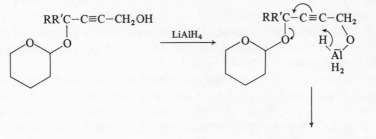

$$RR'C=C=CH-CH_2OH$$

B. NUCLEOPHILIC SUBSTITUTIONS AT UNSATURATED CARBON

3.10. Introduction

Of the many enzymes that catalyse the hydrolysis of peptide linkages, there are a few which require metal ions for their activity. Carboxypeptidase and glycylglycinepeptidase, for example, require Zn^{2+} and Co^{2+} respectively. As the hydrolysis of peptides and of esters and amides of α-amino acids is also catalysed by transition-metal ions, the catalysed reactions are thought to be related mechanistically to the analogous enzymatic processes. Because of this, substitutions at the carbonyl groups of α-amino acid derivatives have featured very strongly among the investigations into the participation of metal complexes in nucleophilic substitutions at unsaturated carbon, with substitutions at aromatic and vinylic centres (see pp. 200 and 208) having received comparatively little attention. The effect of complex formation on nucleophilic substitutions of aromatic halides, however, is clearly a subject that will justify further study, for the few studies that have been made show that the substitutions are enormously enhanced if the aromatic halide is co-ordinated to certain metal carbonyl residues, notably $Cr(CO)_3$ and $[Mn(CO)_3]^+$. At room temperature the reaction of the complex (LXVII; X = Cl) with aniline, for example, is complete within 3 minutes, and affords an excellent yield of the diphenylamine complex (LXVII; X = NHPh). As the $[Mn(CO)_3]^+$ residue is easily displaced by hot acetonitrile, this type of metal-promoted nucleophilic substitution appears to have considerable synthetic potential.

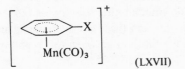

(LXVII)

Substitutions between aryl halides and an extensive range of nucleophiles are also strongly (and uniquely) catalysed by copper and its compounds, and this allows many difficult conversions to be conveniently carried out under laboratory conditions. For example, aryl halides may be converted into the corresponding *N*-aryl phthalimide (and thence by hydrolysis into the primary amine) by the action of copper(I) iodide and potassium phthalimide in refluxing *N,N*-dimethylacetamide. Early work on this type of metal catalysis has been reviewed by Bacon and Hill (1965).

Another type of nucleophilic substitution at unsaturated carbon that is synthetically important and which is currently receiving a great deal of attention (see review by Trost, 1977) is the reaction between π-allylpalladium complexes and nucleophiles such as alcohols, amines, hydride anion, and resonance-stabilised carbanions (figure 3.3).

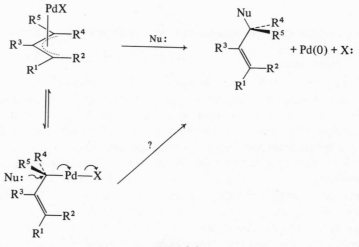

Figure 3.3.

Although this type of substitution is conveniently classified as involving an unsaturated carbon atom, there is a possibility that in some systems the species that is attacked by the nucleophile is a σ-allylpalladium complex. This type of complex is known to exist in equilibrium with the π-allyl form (see figure 3.3) and, of course, contains a metal σ-bonded to a *saturated* carbon atom. Participation of the σ-allyl form is particularly likely when the nucleophile is a resonance-stabilised carbanion, for in this case additional ligands such as triphenylphosphine, 1,2-bis(diphenylphosphine)ethane (DIPHOS), hexamethylphosphoramide and dimethyl sulphoxide (DMSO) have to be added to promote the reaction, and these are the same ligands that have been shown to displace the equilibrium between the π- and σ-forms towards the latter.

One of the reasons for the synthetic importance of this reaction with
π-allylpalladium complexes is the variety of methods available for the preparation of the starting complexes. These include:

(a) reduction of conjugated dienes with palladium(II) chloride in aqueous
acetic acid;

(b) reaction between an alkene and palladium(II) chloride with the latter
in the form of either a $PdCl_2/Na_2CO_3/CH_2Cl_2$ mixture, a $PdCl_2/$
$NaCl/AcOH/AcONa$ mixture, or a complex such as Li_2PdCl_4 or
$Pd(PhCN)_2Cl_2$;

(c) oxidative addition of an allylic chloride, acetate, ether, or alcohol to
a palladium(0) species.

Some of these methods together with subsequent reactions of the resultant
π-complexes with resonance-stabilised carbanions are illustrated in figure 3.4.
As indicated, the products of these reactions are usually consistent with the
nucleophile showing a marked preference to attack the less hindered terminal
carbon atom of the π-allyl system and from that side of the complex that is
opposite to the metal.

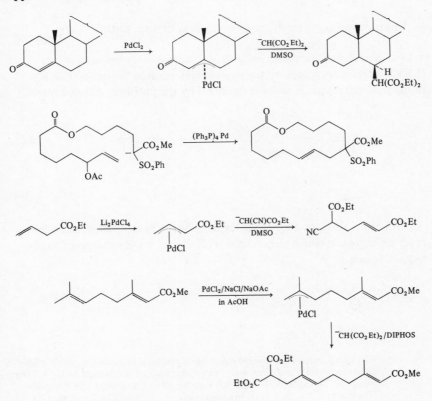

Figure 3.4.

Friedel–Crafts acylations of aromatic compounds have, of course, also been extensively examined. Although these reactions are usually regarded as electrophilic substitutions at unsaturated carbon, they are discussed in this chapter because they are mechanistically related to many other reactions that involve nucleophilic attack at the carbonyl group of a complexed carboxylic acid derivative.

3.11. General comments on metal-catalysed nucleophilic substitutions of carboxylic acid derivatives

In uncatalysed nucleophilic substitutions of carboxylic acid derivatives the nucleophile attacks the carbonyl group and the resultant alkoxide anion expels the leaving group* with the associated pair of bonding electrons:

$$\underset{\underset{O}{\overset{\|}{}}}{\underset{}{R-C-X}} \quad Nu: \longrightarrow \quad \underset{O^-}{\underset{|}{}}{R-\overset{Nu}{\underset{|}{C}}-X} \longrightarrow R-\underset{O}{\overset{\|}{C}}-Nu + X:$$

This process can be catalysed by metal ions in three distinct ways:

1. Co-ordination of the metal ion with the carbonyl oxygen as shown in (LXVIII) facilitates attack by the nucleophile, because the negative charge in the resultant alkoxide anion is stabilised by the positively-charged metal.

$$\underset{\underset{M^{n+}}{\overset{\backslash}{}}}{\underset{O}{\underset{\|}{C}}}{R-C-X^{\cdot}} \quad Nu: \longrightarrow \quad \underset{\underset{M^{(n-1)+}}{\overset{\backslash}{}}}{\underset{:O}{\underset{|}{C}}}{R-\overset{Nu}{\underset{|}{C}}-X} \longrightarrow \underset{\underset{M^{n+}}{\overset{\backslash}{}}}{\underset{O}{\overset{\|}{C}}}{R-C-Nu + X:} \qquad [3.27]$$

(LXVIII)

With esters this type of catalysis leads to acyl–oxygen fission, but a few cases are known in which the complex (LXVIII; X = OR) undergoes alkyl–oxygen fission.

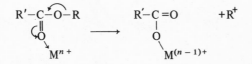

* When the nucleophile is neutral and has a proton attached to its nucleophilic centre, e.g. water and ammonia, this proton may be transferred to the leaving group before the latter is expelled. For the sake of simplicity this complication has been omitted from the discussion, but it should be noted that its inclusion would not radically affect the reasoning that is presented.

The products of this fission are the metal salt of the parent carboxylic acid and the carbonium ion R^+ which reacts with the nucleophile. As in the analogous acid-catalysed process, alkyl–oxygen fission only occurs when the carbonium ion is a comparatively stable one.

2. Metal ions can also catalyse the nucleophilic substitution by co-ordinating with the potential leaving group (X) of the carboxylic acid derivative. The co-ordination increases the polarisation of the C–X bond, and thus facilitates the departure of the leaving group from the intermediate alkoxide anion.

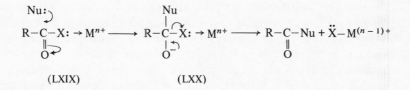

In some systems the polarisation of the C–X bond is so pronounced that heterolytic fission occurs even in the initial complex (LXIX), and this complex then affords the acylium ion (LXXI) at a faster rate than it is attacked by the nucleophile. The fission is followed by a rapid reaction between the acylium ion and the nucleophile.

$$R-C\overset{\frown}{-}X: \to M^{n+} \longrightarrow R-C\equiv\overset{+}{O} \;+\; \overset{..}{X}-M^{(n-1)+}$$

(LXXI)

with Nu: attacking to give

$$R-C-Nu$$
$$\underset{O}{\overset{\|}{}}$$

This heterolytic fission only occurs when very strong Lewis acids and weak nucleophiles are involved, e.g. in the acylation of benzenoid aromatics by acid chlorides and aluminium trichloride.

In addition to facilitating the departure of the leaving group from the reaction intermediate, co-ordination of that group with a metal ion also increases the rate of nucleophilic attack upon the carbonyl group. This is because the bond between the leaving group and the metal involves the lone pair of electrons that participates in the resonance stabilisation of the carboxylic acid derivative (see, for example, (LXXII) and (LXXIII)). The resonance stabilisation of the derivative is therefore decreased by co-ordination, and hence the resonance energy that is lost when the derivative is converted into the alkoxide anion by nucleophilic attack is lower when the derivative is co-ordinated than when it is not.*

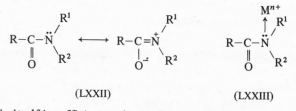

(LXXII)   (LXXIII)

This in itself is sufficient to lower the activation energy required for formation of the alkoxide anion, but superimposed upon this effect is an additional one that also facilitates the reaction. Although the lone pair of electrons on the group X is delocalised to some extent in the carboxylic acid derivative (albeit to a lesser extent when the derivative is co-ordinated), the corresponding lone pair in the alkoxide anion produced by nucleophilic attack is strictly localised. This results in the group X in the anion being a stronger donor, and hence a better ligand for the metal, than the same group in the parent carboxylic acid derivative. From the viewpoint of co-ordination, therefore, the transformation (LXIX) → (LXX) is a favourable process.

3. The third way in which metal ions can promote nucleophilic substitutions of carboxylic acid derivatives is by co-ordinating with both the nucleophile and a functional group (L) in the derivative, to give a complex (LXXIV) whose stereochemistry allows the co-ordinated nucleophile to attack the carbonyl group intramolecularly. Although this type of catalysis has been suggested on many occasions to account for the accelerating effect of metal ions, it has been definitely established for only a limited number of systems.

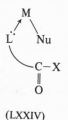

(LXXIV)

Examples of reactions that exhibit the three types of catalysis summarised above are described in the following sections together with related nucleophilic substitutions of compounds other than α-amino acid derivatives.

---

*   It should be noted that the inhibition of the resonance stabilisation of carboxylic acid derivatives R.CO.X that occurs when the group X co-ordinates with a metal ion, is a specific example of a general effect. Other resonance-stabilised compounds in which a lone pair of electrons is directly involved in the stabilisation, e.g. furan and thiophene, will also be destabilised to some extent when the lone pair forms a bond with a metal cation.

3.12. Co-ordination of the carbonyl oxygen of carboxylic acid derivatives

At pH values > 5 the hydrolysis in water of simple α-amino esters proceeds largely by nucleophilic attack by HO^- (rather than H_2O) on the ester group. The rate of attack is accelerated to a small extent when the amino group co-ordinates with a metal ion, because electrostatic effects increase the effective concentration of HO^- in the vicinity of the amino ester. Protonation of the amino group has an identical effect, and the complex $[(NH_3)_5Co \leftarrow NH_2 . CH_2 . CO_2Et]^{3+}$ and the conjugate acid $H_3N^+ . CH_2 . CO_2Et$ react with HO^- at almost the same rate (about 40 times faster than $H_2N . CH_2 . CO_2Et$). In contrast to this comparatively small effect, co-ordination of the ester carbonyl oxygen considerably accelerates the hydrolysis, and the chelated ester in (LXXV; R = Et), for example, reacts with HO^- at about 10^7 times faster than does the free ligand.

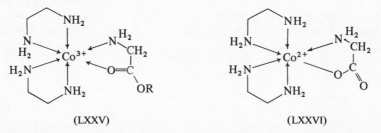

(LXXV)   (LXXVI)

This acceleration due to co-ordination of the ester group was first observed by Kroll (1952) who found that the hydrolysis of α-amino esters in aqueous solution was catalysed by several divalent metal ions, the order of effectiveness being: $Cu^{2+} > Co^{2+} > Mn^{2+} > Ca^{2+}$. Kroll suggested that the catalysis was due to the presence in the solution of the reactive species (LXXVII), in which the α-amino ester was chelated with the metal. This suggestion was supported by a number of subsequent kinetic studies (notably by Bender and Turnquest, and Conley and Martin) which have been summarised by Jones (1968). Particularly relevant to Kroll's suggestion was the discovery that with esters of α-amino acids (such as cysteine, histidine and aspartic acid) which would be expected to chelate with the metal by use of a functional group in the side-chain rather than by use of the ester carbonyl group, the hydrolyses were only slightly catalysed by metal ions. In all cases, the slight increases in rate could be entirely accounted for by the electrostatic effects which would arise from co-ordination of the amino group in the expected chelates (see for example (LXXVIII), p. 162).

One of the disadvantages of all this early work was that on account of the labile nature of the complexes present in the hydrolysis mixtures, it was impossible to isolate and identify the reactive species involved. To avoid this difficulty later workers studied the reactivity of co-ordinated α-amino esters in cations such as $[Co(en)_2(H_2N . CH_2 . CO_2R)]^{3+}$ (LXXV) and the analogous

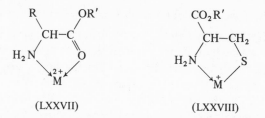

(LXXVII)   (LXXVIII)

triethylenetetra-amine system $[Co(trien)(H_2N \cdot CH_2 \cdot CO_2R)]^{3+}$. These cobalt(III) cations have the advantage that they can be obtained as stable salts which are relatively inert towards ligand-exchange reactions (see p. 50). The infrared absorption spectra of these salts clearly show that in the cations the amino esters are bonded to the metal *via* both their amino and ester groups. The ester group in the tris(perchlorate) salt of (LXXV; R = Et), for example, has a carbonyl stretching frequency of $1625\ cm^{-1}$, a value which is about $115\ cm^{-1}$ lower than that shown by free glycine ethyl ester, and which is consistent with the increase in polarisation caused by co-ordination.

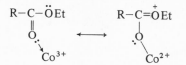

As expected, the co-ordinated ester groups in these cobalt(III) cations are highly activated towards nucleophilic attack. In the range pH = 0 to pH = 4 in which a water molecule is the effective nucleophile, the ester group of (LXXV; R = Et) is hydrolysed at about 10^7 faster than that of the free ligand. The product of the hydrolysis is the glycine salt (LXXVI). Nucleophilic substitutions with amines also proceed at a similar highly enhanced rate; the reactions between the methyl ester (LXXV; R = Me) and α-amino esters to give the corresponding chelated peptide esters are complete within one minute at 20 °C in acetone or dimethylformamide. The use of ^{18}O-labelled ester groups show that although the methyl, ethyl and isopropyl esters are hydrolysed by nucleophilic attack on the carbonyl group followed by displacement of the alkoxy group, the t-butyl ester (LXXV; R = CMe_3) is converted directly into the salt (LXXVI) by alkyl–oxygen fission. In all the hydrolyses the carbonyl oxygen remains bonded to the metal during the course of the reaction.

Activation parameters obtained from kinetic studies of the substitutions with the methyl ester show that the increase in the rate of reaction is entirely due to an increase in the entropy of activation. Presumably this increase is largely associated with the entropy of solvation, for the positive charge on the metal and the negative charge introduced into the co-ordinated ester group by the nucleophile will tend to nullify each other (cf. equation 3.27 on p. 158), and the initial product of the nucleophilic attack will be substantially less solvated than the two individual reactants.

It must be pointed out that although co-ordination of an ester group with a metal ion causes a considerable increase in reactivity towards nucleophiles, this increase is never as great as that caused by protonation. The rate constant for the reaction between water and diprotonated glycine ethyl ester, $H_3N^+ . CH_2 . C(OEt):^+OH$ is about 10^5 greater than that for the analogous reaction with the cation (LXXV; R = Et). In addition, although activation of an ester group by co-ordination has the advantage that it can be accomplished over a wide pH range, it has the disadvantage that in order for it to occur in the first place, the carboxylic ester group must be adjacent to another donor centre. This is because the ester group is a comparatively weak ligand, and for co-ordination to occur to an appreciable extent (particularly in aqueous solution), the ester group has to be part of a chelating system. Adjacent amino, amido and carboxylate groups are effective in this context. Even a hydroxyl group is effective to some extent, as shown by the relative effects caused by metal ions (e.g. La^{3+}, Ce^{3+}) on the rates of hydrolysis of ethyl mandelate and ethyl phenylacetate. An adjacent alkoxy group is particularly effective, and the direct preparation of β-hydroxy acids by use of t-butyl α-bromoesters in Reformatskii reactions has been rationalised in terms of alkyl–oxygen fission of a co-ordinated ester group.

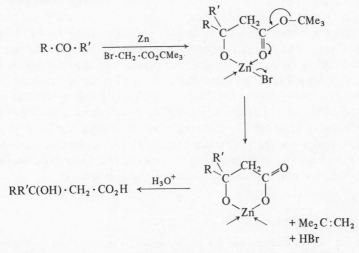

Investigations into the hydrolysis of various types of amides show that an amide group is also activated towards nucleophilic attack when the carbonyl oxygen is co-ordinated with a metal ion. The structural requirement for effective co-ordination (and hence activation) is the same as with carboxylic esters, i.e. the carbonyl oxygen must form part of a chelating system. Amido, amino and carboxylate groups can all function as the other co-ordinating centre(s) of the chelate system, and consequently the hydrolyses of α-amino amides, α-acylamino acids, and di-, tri- and higher peptides are catalysed by metal ions. At pH = 9, for example, the hydrolysis of glycine amide is catalysed by Cu^{2+}, Co^{2+} and Ni^{2+}, with the first ion being the most effective. In the absence of metal ions, phenylalanylglycine amide (LXXIX) cyclises in aqueous solution to give 3-benzyl-2,5-diketopiperazine (LXXX) by an intramolecular reaction between the terminal $-NH_2$ and $-CO.NH_2$ groups. In the presence of Cu^{2+}, however, the hydrolyses of the peptide and the amide links are competitive with the cyclisation.

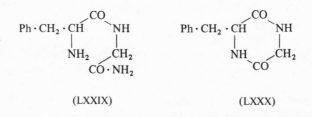

(LXXIX)   (LXXX)

A limited amount of quantitative information on the activation of amide groups towards hydrolysis has been obtained by use of the $[Co(en)_2]^{3+}$ and $[Co(trien)]^{3+}$ ions (cf. p. 161). A number of salts are known in which these ions are chelated with an α-amino amide, a peptide, or a peptide ester. The spectroscopic properties, and in some cases the X-ray structures, of these salts show that in the cation the organic ligand is chelated with the metal through the terminal NH_2 group and the carbonyl oxygen of the adjacent amide group, e.g. the cation $[Co(en)_2,gly.NMe_2]^{3+}$ has the structure (LXXV; replace OR by NMe_2). A comparison of the relevant rate constants shows that the co-ordinated amide groups in these salts are hydrolysed at about 10^6 times faster than the corresponding group in the free ligand. Although this rate enhancement is quite dramatic, it is still up to 10^4 times smaller than that which results when metalloenzymes are used to catalyse the hydrolysis of amino amides. Clearly the increased reactivity of co-ordinated amide groups is not sufficient by itself to explain the rapidity of the enzyme-catalysed reactions, and other activating effects must be involved.

3.13. Co-ordination of the leaving group

For the reasons discussed in §3.11, nucleophilic substitutions at the carbonyl group of a carboxylic acid derivative R.CO.X are accelerated when the leaving group X co-ordinates with a metal ion. This acceleration can be utilised in organic synthesis when a carboxylic acid derivative is to react with a weak nucleophile, and the most frequent use is in Friedel–Crafts acylations. In these reactions an aromatic compound is treated with a mixture of a Lewis acid such as $AlCl_3$, BF_3, $SbCl_5$ or $FeCl_3$, and a carboxylic acid derivative. Acid chlorides or anhydrides are generally used as they are more reactive than esters and amides. Under the influence of the Lewis acid, the aromatic

compound and the carboxylic acid derivative react to give an acylcyclo-
hexadienyl cation which loses a proton to form the acylated product.

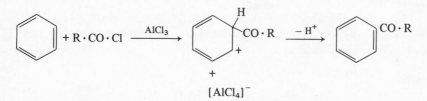

[AlCl₄]⁻

In the case of an acid chloride, co-ordination with the Lewis acid can
involve either the halogen (see LXXXII) or the carbonyl oxygen (LXXXI).
Those complexes of the latter type, e.g. Et . CO . Cl/SbCl₅, are characterised
by an increased carbon–oxygen bond length and a downfield shift of τ0.5–1.0
of the α-protons. Although such complexes can probably function as acylating
agents (see p. 161), it is widely believed that in Friedel–Crafts acylations
(particularly with relatively unreactive substrates) the acylating agent is the
acylium ion (LXXXIII) formed by heterolytic fission of the carbon–chlorine
bond in a complex of type (LXXXII). Many Lewis acid/acyl halide combinations afford acylium ions in the form of stable salts, e.g. (Me . CO)⁺ (BF₄)⁻,
and the large downfield shift (τ1.0–2.0) of the α-protons and the marked
deshielding of the carbonyl carbon in the ¹³C n.m.r. spectrum indicate that
the positive charge of the acylium cation residues largely on carbon. As
expected these acylium salts are effective acylating agents, but as cryoscopic
measurements show that even in highly polar media the salts undergo only a
low degree of ion separation it is possible that the active acylating species is
an ion-pair.

$$
\begin{array}{ccc}
\underset{\displaystyle X}{\overset{\displaystyle O \rightarrow M^{n+}}{R-C}} & \rightleftharpoons & \underset{\displaystyle X \rightarrow M^{n+}}{\overset{\displaystyle O}{R-C}} & \rightleftharpoons & [R-C\equiv\overset{+}{O} \leftrightarrow R-\overset{+}{C}=O] \\
& & & & \text{(LXXXIII)} \quad + \quad X-M^{(n-1)+}
\end{array}
$$

(LXXXI)   (LXXXII)

As an acylium ion (R . CO)⁺ can undergo decarbonylation to give the
corresponding carbonium ion (R⁺), the acylation of an aromatic compound
under Friedel–Crafts conditions is often accompanied by alkylation. The
decarbonylation is favoured when R⁺ is a stable carbonium ion, and consequently alkylation is particularly observed in the use of the acid halides of
tertiary and α-aryl carboxylic acids.

$$Ph \cdot H + Me_3C \cdot CO \cdot Cl \xrightarrow{AlCl_3} Ph \cdot CMe_3 + HCl + CO$$

$$Ph \cdot H + Ph \cdot CHMe \cdot CO \cdot Cl \xrightarrow{AlCl_3} Ph \cdot CH(Me)Ph + HCl + CO$$

The detailed mechanism, the scope, and other aspects of Friedel–Crafts acylations have been discussed at length in a series of volumes edited by Olah (1963).

Mechanistically related to Friedel–Crafts acylations are the Fries rearrangements which aryl esters undergo when treated with a strong Lewis acid such as aluminium trichloride.

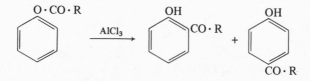

The essential difference between the two types of reaction is that in a Friedel–Crafts reaction the nucleophile which reacts with the acylium ion is the aromatic compound to be acylated, but in a Fries rearrangement the nucleophile is the gegenion produced with the acylium ion from the initial complex.

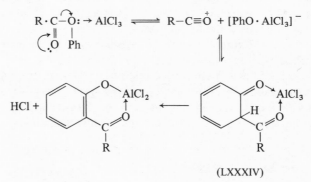

(LXXXIV)

The acylium ion can react with the gegenion either at the *ortho*- or the *para*-position of the aryl residue in the latter species. Reaction at the *ortho*-position is often favoured, probably because in this case the aluminium salt is stabilised by chelation as shown in (LXXXIV). Analyses of the products obtained when mixtures of aryl esters are subjected to the Fries rearrangement show that the reaction between the acylium ion and its gegen ion involves free solvated ions, as well as ion-pairs.

As both Friedel–Crafts acylations and Fries rearrangements require the CO—X bond in the co-ordinated carboxylic acid derivative to be sufficiently polarised for acylium ion formation to occur, weaker Lewis acids are ineffective at promoting these types of reaction. When co-ordinated with the leaving group, however, they do accelerate the rate at which the carbonyl group in the carboxylic acid derivative is attacked by nucleophiles. The simplest yet most convincing demonstration of this is provided by the strong catalytic effect which the heavy metal ions Hg^{2+}, Pb^{2+} and Ag^+ exert on the hydrolysis, aminolysis and ester-exchange reactions of thiol esters. In the presence of mercury(II) acetate, for example, the reaction between ethyl thioacetate and ethanol is almost instantaneous at room temperature.

$$Hg(OAc)_2 \rightleftharpoons \overset{+}{Hg}\cdot OAc + AcO^-$$

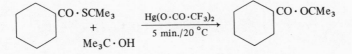

Indeed, even an ester-exchange reaction between a t-butyl thiol ester and t-butyl alcohol can be made to proceed at room temperature by use of the more reactive mercury(II) trifluoroacetate (cf. p. 179).

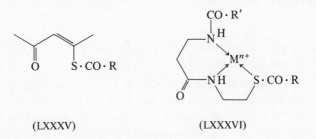

Metal ions such as Cu^{2+}, Ni^{2+} and Co^{2+} which have lower affinity for sulphur are not so effective at promoting these nucleophilic substitutions, except in the case of thiol esters in which the $-SR$ group contains additional donor centres, e.g. as shown in (LXXXV). With these esters the comparatively low affinity of the sulphur for the metal ion is overcome by the fact that the sulphur atom constitutes part of a chelating system (cf. co-ordination of the carbonyl oxygen, p. 163). The importance of additional donor centres in thiol esters may be of significance in the roles of coenzyme-A esters as biological acylating agents. The susceptibility towards nucleophilic attack by alcohols and amines, for example, would certainly be substantially increased if the esters were to act as tridentate chelating agents (see LXXXVI).

(LXXXV)   (LXXXVI)

Nucleophilic substitutions of carboxylic esters are similar to those of thiol esters in that the $-OR$ group must be able to function as a chelating system for the catalysis by transition-metal ions to be of significance.

Thus although Cu^{2+} and Ni^{2+} have very little effect on the hydrolysis and ester-exchange reactions of simple esters, they strongly catalyse the reactions

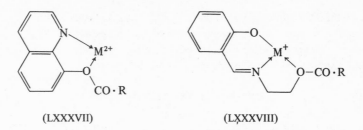

(LXXXVII)   (LXXXVIII)

of esters of 8-hydroxyquinoline and *N*-(salicylidene)-2-aminoethanol because
of the formation of chelates of types (LXXXVII) and (LXXXVIII) respectively.

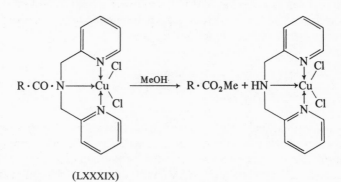

(LXXXIX)

Similarly, the metal-ion catalysis of nucleophilic substitutions of amides is
most pronounced when the amide nitrogen is part of a chelating ligand. A
most convincing demonstration of this is provided by the extraordinary
rapidity with which copper(II) chloride complexes (LXXXIX) of *N,N*-di-
(2-pyridylmethyl)amides undergo methanolysis. Even with the sterically
hindered complex (LXXXIX; R = t-Bu) the reaction is complete in boiling
methanol within five minutes.

3.14. Nucleophilic substitutions at the β-position of αβ-unsaturated ketones and related systems

In the metal–catalysed reactions described in §3.12 the metal facilitates
nucleophilic substitutions at the carbonyl group by stabilising the negative
charge generated on the carbonyl oxygen by nucleophilic attack. Analogously,
nucleophilic substitutions at the β-position of αβ-unsaturated carbonyl compounds are also facilitated if the carbonyl oxygen is co-ordinated with a metal
ion, for the metal can stabilise the intermediate carbanion (see p. 169).

An example of this effect is provided by the transformation (XC)→(XCI)
which readily takes place when the nitrogen atom and the adjacent carbonyl
group of (XC) are chelated with a transition-metal ion. Similarly the

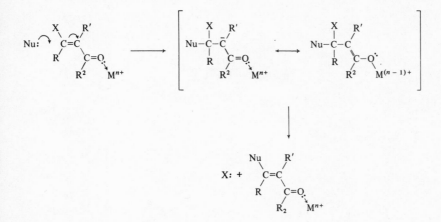

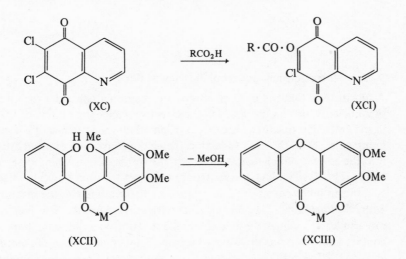

(XC)   (XCI)

(XCII)   (XCIII)

dibenzpyran (XCIII) is formed more rapidly from the diaryl ketone (XCII; M = H) when the latter is in the form of a metal complex (e.g. XCII; M = Zn^+).

A very interesting result which is observed when thallium(III) acetate is used for promoting nucleophilic attack by methanol on β-aryl-$\alpha\beta$-unsaturated ketones, is that in the absence of a leaving group at the β-position the initial product undergoes further nucleophilic attack and a 1,2-shift of the aryl group. Thus thallium(III) acetate in methanol converts chalcones of the type (XCIV) into the acetals (XCV) (see p. 170).

These reactions constitute part of a useful synthesis of isoflavones from chalcones, for if the substituent R in (XCV) is a benzyloxy group, removal of the benzyl group followed by acid-catalysed hydrolysis of the acetal group causes spontaneous cyclisation to give the isoflavone (XCVI).

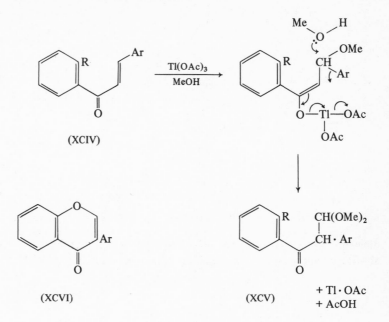

(XCIV)

(XCVI)

(XCV)

+ Tl · OAc
+ AcOH

3.15. Intramolecular nucleophilic substitutions

A substitution that occurs at an unsaturated carbon atom under the influence
of a nucleophilic reagent in which the effective nucleophile is one of the
ligands attached to a metal is often accelerated by the presence of a donor
group adjacent to the site of reaction. Thus, the conversion of ethyl esters
into the corresponding benzylamides by the action of benzylaminomagnesium
iodide ($PhCH_2NH . MgI$) proceeds more readily if the ester group is adjacent
to a substituent with an oxygen or nitrogen donor atom, e.g. the reaction
with ethyl picolinate is faster than that with ethyl benzoate. This increase in
rate can lead to selectivity in nucleophilic substitutions with compounds that
contain more than one unsaturated centre. For example, when the diethyl
ester (XCVII; R = Et) is heated with one equivalent of magnesium methoxide,
the ester-exchange reaction of the ester group *ortho* to the hydroxyl group is
sufficiently faster than that of the other ester group to ensure that the
product of the reaction is largely the mixed ester (XCVII; R = Me). Similarly,
although molecular models show little difference in the accessibility of the
two ester groups in the morphine analogue (XCVIII), the reaction with
methylmagnesium iodide preferentially involves the group at C_7, i.e. the one
next to a methoxyl group.

This ability of donor groups to induce selectivity is shown by the terminal
$-NH_2$ group of simple peptides, in that this group can selectively activate
the hydrolysis of the *N*-terminal amide linkage by certain metal complexes,

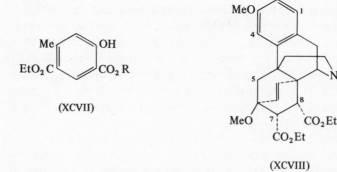

(XCVII)

(XCVIII)

notably the cation *cis*-hydroxyaquotriethylenetetraminecobalt(III), [Co-trien(OH)(H$_2$O)]$^{2+}$. Treatment of this cation with tetraglycine, for example, affords quantitative yields of triglycine and the glycine salt [Co-trien(gly)]$^{2+}$. This type of reaction could, of course, be used for the determination of the *N*-terminal residue of a simple peptide, but its importance lies more in its great similarity to the selective hydrolysis induced by certain metalloenzymes.

The effect of a donor group on the rate of nucleophilic substitution at a carbonyl group can be easily explained by assuming that co-ordination of the donor group with the metal affords a complex of the general type (LXXIV) shown on p. 160. The nucleophilic substitution can then proceed by a comparatively fast intramolecular route. The validity of this explanation, together with a quantitative assessment of the increase in rate associated with this intra-molecular route, has been tested in several specific cases by use of the substitutionally-inert cobalt(III) ion (see also pp. 161 and 164).

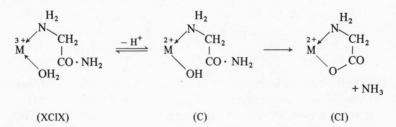

(XCIX)   (C)   (CI)

The complex cation *cis*-[Co(en)$_2$Br(gly . NH$_2$)]$^{2+}$, for example, has been used to determine the rate of nucleophilic attack by HO$^-$ on glycine amide when both these reactants are adjacent to each other in the co-ordination sphere of Co^{3+}. In aqueous solution at pH = 9, this complex forms the corresponding aquo species (XCIX; M = Co(en)$_2$), the co-ordinated water molecule of which is fairly readily deprotonated because of the increased acidity caused by co-ordination (see p. 106). In the deprotonated complex (C) the co-ordinated hydroxyl group rapidly attacks the glycine amide to afford the glycine chelate [Co(en)$_2$(gly)]$^{2+}$, i.e. (CI; M = Co(en)$_2$). The rate data for the various stages of the hydrolysis reveal that this cobalt(III)-induced hydrolysis of glycine amide is at least 10^7, and possibly more than 10^{11}, times faster than the uncatalysed hydrolysis of the free amide. A similar dramatic rate-enhancement has been observed in the cobalt(III)-induced hydrolysis of glycine ethyl ester. Evidently the intramolecular nature of the nucleophilic substitution involved in these hydrolyses more than compensates for the greatly reduced basicity (and presumably nucleophilicity) of co-ordinated HO^-.

The results obtained with these cobalt(III) systems not only confirm that nucleophilic substitutions can proceed by the intramolecular route, but they also demonstrate that as far as amino acid derivatives are concerned this route is decidedly more efficient than the intermolecular one which involves coordination of the carbonyl oxygen (see p. 161).

3.16. Directive effects in metalation

One of the substitution processes exhibited by organometallic compounds is metal–hydrogen exchange:

$$R.M + R'.H \rightleftharpoons R.H + R'.M \qquad [3.28]$$

This process, commonly termed metalation, is frequently used for preparing organometallic compounds that cannot be prepared by the more direct route of halogen–metal exchange, as in the following examples.

$$Ph \cdot C \equiv C \cdot H \quad + Me \cdot MgX \longrightarrow Ph \cdot C \equiv C \cdot MgX + Me \cdot H$$

On the basis of large primary isotope effects and comparisons between metalation and base-catalysed hydrogen–deuterium exchange, it is considered that metalations with organometallics of the alkali and alkaline earth metals involve abstraction of a proton from the organic substrate, possibly *via* a four-membered transition state (see (CII)).

(CII)

It should be noted that this type of metalation is mechanistically different from that discussed in §3.18, which is an electrophilic process.

The equilibrium position of reaction 3.28 is determined by the relative acidities of the two hydrocarbons RH and R'H, and lies on the side of the weaker acid. The metalating ability of the organometallic (RM) is therefore directly related to the basicity of the carbanion R^-, and decreases in the order $-\overset{|}{\underset{|}{C}}{}^- > \overset{}{\underset{}{C}} = C^- > -C \equiv C^-$. For the same reason, the effectiveness of simple metal alkyls as metalating agents increases as the alkyl residue is changed from primary to secondary to tertiary.

The actual site of metalation of an organic compound cannot always be predicted from a consideration of the relative acidities of the various hydrogen atoms, for a high proportion of metalations are subject to kinetic rather than thermodynamic control. In such cases, the product is determined by the nature of the metalating agent which is used, as well as by the reaction conditions. With some systems the kinetically-controlled product may subsequently undergo conversion into the thermodynamically stable product, but when this process occurs it is always slower than the initial metalation.

In the reaction between isopropylbenzene and pentylpotassium (see table 3.1), for example, the initial product contains a mixture of *m*- and *p*-isopropylphenylpotassium formed in a kinetically-controlled process. In the presence of an excess of isopropylbenzene, both these compounds are converted by trans-metalation into α,α-dimethylbenzylpotassium, i.e. the more stable isomer. Metalation with n-butyl-lithium/tetramethylethylenediamine also gives a mixture of organometallics whose composition is kinetically controlled, but in this case the composition of the mixture shows hardly any variation with reaction time.

Table 3.1. *Metallation of isopropylbenzene (from Benkeser et al., 1963)*

Metallating agent	Time (h)	Position of metallation and product composition (per cent)			
		Side-chain	*ortho*	*meta*	*para*
n-C$_5$H$_{11}$K	3	42	–	39	18
	20	88	–	4	8
	48	100	–	–	–
n-BuLi/TMEDA	2	3	10	57	30
	24	3	8	59	30

Kinetic control of the site of metalation is of particular importance in the metalation of organic substrates that contain co-ordinating groups, for by co-ordinating with the organometallic reagent such groups can considerably increase the rate at which an adjacent position is metalated. This increase in rate is often large enough to ensure that the organic substrate is metalated selectively, if not specifically, at the adjacent position. The methoxyl group in anisole for example, activates the *ortho*-position sufficiently to allow this compound to be metalated at a reasonable rate by Grignard reagents; under the same reaction conditions benzene is completely unchanged.

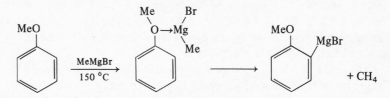

Most of the recent studies on this activating effect have concerned the metallation by organolithium reagents of aromatic compounds that contain ether and tertiary amine groups. Shown below are the structures of some of these compounds, and the positions at which they are metalated. For a useful review on the earlier work on hydrogen–lithium exchange and information on other directing groups, the reader is referred to the reviews by Gilman and Morton (1954), and Bruce (1977).

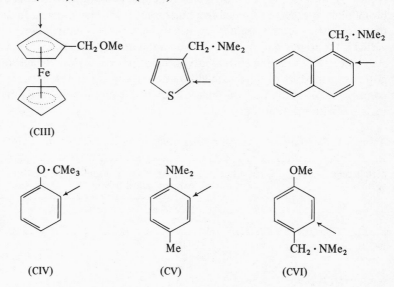

The very strong tendency for metalation to occur at positions adjacent to a co-ordinating group is particularly manifested in the reactions of t-butyl-phenyl ether (CIV) and *p*-dimethylaminotoluene (CV). With n-butyl-lithium in ether, both these compounds are lithiated exclusively (> 99 per cent) at the *ortho*-position, in spite of the substantial steric hindrance in the ether and the presence of the more acidic hydrogens of the aromatic methyl group in the amine. The exceptionally strong directing effect of a $-CH_2NMe_2$ group is revealed by the metalation of the amino ether (CVI).

The metalation by organolithium reagents of compounds that contain a —CH$_2$OR group attached to a benzene ring causes exchange of the benzylic hydrogens. The initial product, however, undergoes a Wittig rearrangement to give a lithium alkoxide:

$$\text{Ph} \cdot \text{CH}_2 \cdot \text{OMe} \rightarrow \underset{\underset{\displaystyle \text{Li}}{|}}{\text{Ph} \cdot \text{CH} \cdot \text{OMe}} \rightarrow \text{Ph} \cdot \text{CH(Me)} \cdot \text{OLi}$$

In contrast to this result, the metalations of the corresponding ferrocene systems afford the nuclear substituted organolithium derivative (see (CIII)). This is because the acidity of the hydrogen atoms of a CH$_2$ group attached to a ferrocene system is very much lower than that of benzylic hydrogen atoms.

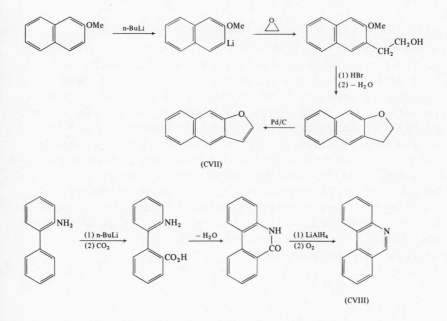

(CVII)

(CVIII)

Selective metalations have often proved to be extremely useful in organic synthesis. For example, convenient syntheses of naphtho [2,3-*b*] furan (CVII) and phenanthridine (CVIII) from 2-methoxynaphthalene and 2-aminobiphenyl respectively, utilise the fact that lithiation of the two starting materials results in the introduction of lithium at the correct position for the subsequent annellation reactions.

C. ELECTROPHILIC SUBSTITUTIONS AT UNSATURATED CARBON

3.17 Introduction

The major portion of the published work on metal complexes in electrophilic substitutions at unsaturated carbon concerns the Friedel–Crafts acylation of aromatic compounds. As this particular topic has been reviewed on many occasions - notably by Olah (1963) - and is adequately dealt with by most standard texts on organic chemistry, it is only briefly discussed in this book and for convenience the discussion is included in §3.13. In this section the two topics that are discussed concern reactions between (*a*) electrophilic reagents and metal complexes of certain organic ligands, and (*b*) metal ions and aromatic compounds.

3.18. Electrophilic substitutions of co-ordinated organic ligands

Aromatic rings that bear a primary, secondary, or tertiary amino group are usually highly reactive in electrophilic substitutions because the lone pair of electrons on the nitrogen atom facilitates the introduction of a positive charge into the ring during the rate-determining attack by the electrophile. This activating effect is lost when the amino group is protonated, for the lone pair of electrons is then directly involved in bond formation, and is no longer available for stabilisation of the intermediate carbonium ion. The conjugate acid of the amine is therefore much less reactive towards electrophilic attack. As the lone pair of electrons is also involved when the amino group co-ordinates with a metal ion, co-ordination similarly deactivates the aromatic ring of the amine towards attack by electrophiles. However, kinetic studies have revealed that this deactivation is not as great as that caused by protonation, and that even when co-ordination to a triply-charged metal ion is concerned the order of reactivity is still:

$$
\underset{\substack{\downarrow \\ M^{n+}}}{\overset{R}{\underset{|}{Ar-\underset{\cdot\cdot}{N}-R'}}} \; > \; \underset{\substack{\downarrow \\ M^{n+}}}{\overset{R}{\underset{|}{Ar-\underset{|}{N}-R'}}} \; > \; \underset{\substack{| \\ H}}{\overset{R}{\underset{|}{Ar-\overset{+}{N}-R'}}}
$$

This also applies to the relative effect of co-ordination and protonation on the reactivity of phenolate anions towards electrophilic reagents, and the order of reactivity which is observed is: $ArO^- > ArO-M^{n+} > ArO-H$.

In the few unambiguous cases that have been examined to date, co-ordinated aromatic amines, heterocyclic amines (e.g. pyridine) and phenolate anions have been found to undergo electrophilic substitutions at the same position in the aromatic ring as the free ligands. This situation is not unexpected with the last two types of compound, but considerations based on the

apparent similarity between protonation and co-ordination ((CIX) and (CX)
respectively) suggest that a co-ordinated amino group should behave like a
$-^+NHR_1R_2$ group and be *meta*-directing rather than *ortho* and *para* as is
actually found. Evidently co-ordination does not immobilise the nitrogen
lone pair as efficiently as protonation does, and the lone pair can still participate in the mesomeric effect to some extent when the aromatic amine is
attacked by electrophiles (see also p. 130).

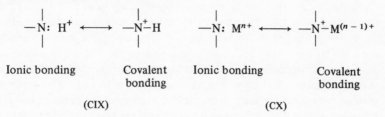

Ionic bonding	Covalent bonding	Ionic bonding	Covalent bonding
(CIX)		(CX)	

The attachment of an organic ligand to a metal ion often affords a complex that can be subjected to reactions under conditions in which the free
ligand would either be rapidly decomposed or be incapable of existence. A
good example of this is the acetylacetonate anion which when chelated to
transition-metal ions such as Cr^{3+}, Co^{3+} and Rh^{3+}, can be made to undergo
electrophilic substitutions at the 3-position.

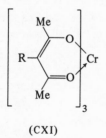

(CXI)

For example, by use of suitable reagents tris(acetylacetonato)chromium(III)
(CXI; R = H) has been halogenated, nitrated, formylated, chloromethylated and
acetylated to give trisubstituted chelates of type (CXI). These and other substitutions and the reactions of the substituted chelates have been discussed by
Collman (1965). One particular point of interest is that the introduction of
an electron-withdrawing substituent, e.g. NO_2 or COMe, into one of the acetylacetonate ligands deactivates the other two towards electrophilic attack, and so
the three substitutions which lead to the final trisubstituted product (CXI)
take place progressively less readily.

The similarity between these electrophilic substitutions and those of
benzenoid aromatics has lead to the suggestion that the metal acetylacetonates are aromatic compounds. However, this suggestion is refuted by

n.m.r. spectroscopy which shows that the chelate rings do not sustain an induced ring current, and hence are not aromatic in the modern sense of the word.

For the organic chemist the preparation of metal chelates of 3-substituted acetylacetone is of limited synthetic use. In most cases those β-diketone ligands that cannot be made by conventional methods cannot be isolated from their metal chelates, because under the conditions required to remove the metal from the chelate the free ligand is rapidly decomposed. Fortunately this situation does not apply to all chelate systems, and as mentioned earlier (p. 92) organic ligands can often be easily isolated from an iron tricarbonyl complex by treatment with a mild oxidising agent. This is fortunate, for there are a number of olefinic ligands that do not normally exhibit electrophilic substitutions but that do so when co-ordinated with an iron tricarbonyl residue. For example, with electrophilic reagents cyclo-octatetraene gives products arising from rearrangements of the eight-membered ring, but when the complex (CXII; R = H) is acetylated (AlCl$_3$/MeCOCl) and formylated (POCl$_3$/H . CONMe$_2$) the monosubstituted derivatives (CXII, R = COMe and CHO respectively) are obtained. Removal of the iron tricarbonyl residues from these complexes by treatment with cerium(IV) ions affords the corresponding monosubstituted cyclo-octatetraene. The highly reactive compound tetraiodocyclobutadiene has similarly been obtained from cyclobutadienyliron tricarbonyl (CXIII; R = H) (see p. 94) by treating this complex with an excess of mercury(II) acetate to give (CXIII; R = HgOAc), and this with KI$_3$ gave (CXIII; R = I) from which the tetraiodocyclobutadiene was liberated by the action of cerium(IV) ions.

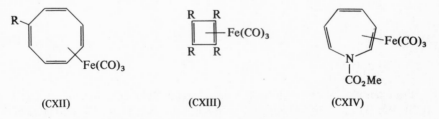

(CXII)   (CXIII)   (CXIV)

Other organic ligands that are attacked by electrophiles when co-ordinated to iron tricarbonyl include cyclohepta-1,3,5-triene and buta-1,3-diene (both are acylated at position 1), norbornadiene (formylation at position 1), and *N*-methoxycarbonylazepine (see (CXIV); acylation at position 3).

The fact that certain co-ordinated olefinic ligands react with electrophiles to give carbonium ions which 'revert to type' by expelling a proton rather than acquiring a nucleophile to give an addition product, clearly indicates the high thermodynamic stability of the initial type of complex.

Co-ordination to metal carbonyl residues also modifies the reactivities of benzenoid aromatics towards electrophilic substitutions. In Friedel–

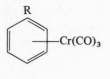

(CXV)

Crafts acylations of the chromium tricarbonyl complexes (CXV) of mono-substituted benzenes, for example, the usual activating effect on the *para*-position by alkyl and alkoxy groups is largely lost, and acetylation of the t-butyl complex (CXV; R = t-Bu) gives the *meta*- and *para*-disubstituted complexes in the ratio 87 : 13. This contrasts with the free ligand which under the same conditions gives a 4 : 96 ratio.

3.19. Metal cations as electrophiles

The electrophilic displacement of a proton from a benzenoid or heterocyclic compound by a metal ion is largely confined to ions with class '*b*' character, of which Hg^{2+}, Pd^{2+}, Pt^{2+} and Tl^{3+} are the most important in this context. These ions are used in the form of salts which dissociate to some extent in the reaction media, and hence the substitutions are facilitated by polar solvents and salts that are extensively dissociated in solution. In mercurations, for example, mercury(II) trifluoroacetate is far more reactive than the acetate which in turn is more reactive than the highly covalent mercury(II) chloride:

$$(CF_3 . CO_2)_2 \, Hg \rightleftharpoons CF_3 . CO_2 Hg^+ + CF_3 . CO_2^-$$

$$CF_3 . CO_2 Hg^+ + ArH \rightleftharpoons CF_3 . CO_2 Hg . Ar + H^+$$

Substitutions of aromatic compounds by metal ions are directly analogous to other electrophilic substitutions, and substituents already present in the ring usually confer the expected pattern of reactivity upon the different positions in the ring. Exceptions arise, however, with substituents that can co-ordinate with the metal ion, e.g. $-CH_2NMe_2$, $-N=N-$ and $-(2\text{-pyridyl})$, and these substituents exert a very pronounced *ortho*-directing effect. Thus only the *ortho*-substituted derivatives (CXVI; R = CH_2OH and CO_2Me) are obtained by the action of thallium(III) trifluoroacetate on the appropriate mono-substituted benzene because of the directing effect of the oxygen-containing substituent. Similarly 2-phenylpyridine and azobenzene both react with Na_2PdCl_4 to give almost exclusively the *ortho*-substituted products (CXVII) and (CXVIII) respectively as halogen-bridged dimers (see p. 3). In the case of the last reactions, however, it should be mentioned that with substitutions by palladium(II) – and indeed with most substitutions by transition-metal ions – the actual mechanisms of the reactions are still uncertain and it is possible that they are initiated by oxidative addition of an aromatic C—H bond to the metal (see p. 229) rather than by electrophilic attack by a metal cation.

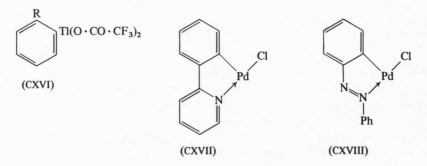

(CXVI)   (CXVII)   (CXVIII)

As a result of the nature of the metal involved, the organometallic compounds produced by these substitutions are comparatively poor bases and nucleophiles, and their synthetic uses are different from those of the organolithium compounds, for example. Replacement of the metal by deuterium and by halogens are the two commonest reactions, and these two processes are carried out by use of suitable reducing agents such as lithium aluminium deuteride or deuterium gas, and by direct action with halogens or (in the case of thallium(III) compounds) halide anions respectively. Mercuration followed by treatment of the resultant organo-mercury compound with iodine is a very convenient method for introducing this halogen into reactive aromatic compounds.

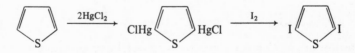

Electrophilic substitution by a metal ion followed by replacement of the metal constitutes a useful route for the selective introduction of deuterium or a halogen into the *ortho*-position of an aromatic ring that bears one of the substituents mentioned above. Thus, when treated with potassium iodide the thallium compound (CXVI; R = CO_2Me) obtained from methyl benzoate gives methyl o-iodobenzoate. Similarly reduction of (CXVII) with lithium aluminium deuteride gives 2-(2-^{2}H-phenyl)pyridine.

4   Metal complexes in additions

4.1.   Introduction

In the absence of metal ions, additions to multiple bonds between carbon
and oxygen or nitrogen can be initiated either by direct nucleophilic attack,
or by nucleophilic attack after prior protonation of the oxygen or nitrogen.

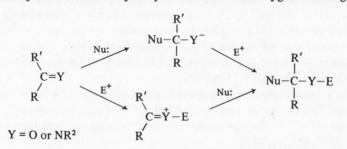

With the exception of compounds in which the double or triple bond is
attached to substituents that can stabilise a negative charge, e.g. F or CO_2R,
this prior combination with a proton or an equivalent electrophilic species
(E^+) is invariably required for additions to carbon–carbon double and triple
bonds, and affords a positively-charged species usually represented as a classical carbonium ion or a π-complex, e.g. (I) or (II).

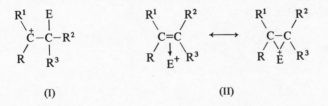

Additions to multiple bonds can also be initiated by co-ordination with a
metal ion for, as with protonation, co-ordination results in the formation of a
species that is far more susceptible to nucleophilic attack than the original

substrate. The analogy between protonation and co-ordination may be extended further, for in most metal-ion promoted additions to multiple bonds between carbon and oxygen or nitrogen the metal is co-ordinated with the oxygen or the nitrogen, but in additions to alkenes and alkynes the reactions involve π-complexes.

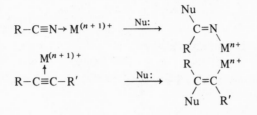

In these reactions, nucleophilic attack on the co-ordinated substrate leads initially to a species in which the metal is linked to the organic ligand by a σ-bond. In some systems this species is stable enough to be isolated, and is therefore the actual addition product. In others, protolysis of the metal-ligand σ-bond gives rise to a metal-free addition product. Occasionally the species is unstable and undergoes a rearrangement which results in oxidation of the ligand and reduction of the metal. Although in such cases the overall reaction is not one of addition, this type of process is discussed in this chapter (§4.6) for reasons of convenience.

One general consequence of this initial formation of a σ-bonded species is that the ability to initiate additions to alkenes and alkynes is largely restricted to those metal ions that form relatively stable σ-bonds with carbon, e.g. Hg^{2+}, Pd^{2+}, Pt^{2+} and Tl^{3+}.

As in the nucleophilic substitutions discussed in chapter 3, the reaction between a co-ordinated substrate and a nucleophile can be either intermolecular, or intramolecular if the nucleophilic centre is in one of the ligands co-ordinated with the same metal ion as the substrate. With intramolecular additions in which the nucleophile is a ligand co-ordinated with the metal through its nucleophilic centre, as with HO^-, and the substrate is an alkene, alkyne, or some other π-bonded ligand, then the addition constitutes an example of an insertion.

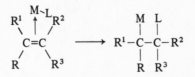

One example of this situation is given in §4.5, while others in which the nucleophilic ligand is a carbanion or hydride anion are discussed under the general heading of Insertion in §5.3.

In the reactions summarised above, a nucleophile attacks an unsaturated ligand that is bonded to a metal *via* an atom on which negative charge accumulates as the result of the nucleophilic attack, i.e. the metal is bonded to this particular atom *before* attack occurs. There are, however, a few examples of reactions where co-ordination accompanies nucleophilic attack, for example in reactions of the general type:

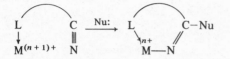

Specific examples of this situation are given on p. 185.

There are also examples of additions in which metal complexes are involved where the role of the metal ion is largely stereochemical, rather than electronic as in those described above. For example, although non-activated double and triple bonds do not normally add Grignard reagents or alkyl-lithium compounds, the presence of an adjacent co-ordinating substituent (OMe, NMe_2) allows intramolecular addition to take place *via* a metal complex in which the organometallic reagent is held close to the unsaturated centre.

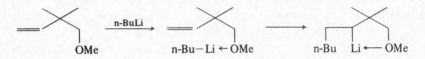

A final point is that in a number of additions to carbon–carbon and carbon–oxygen double bonds, stereoselectivity can arise if the reaction mixture contains chiral ligands. Specific examples of this situation are given in §4.8.

In the remaining sections of this chapter the various aspects of metal-promoted additions indicated above are discussed in detail.

4.2. Additions to carbon–nitrogen triple bonds

With very few exceptions the complexes formed from metal ions and organic cyanides involve σ-bonding with the nitrogen as the donor, and the positive charge can be regarded as being delocalised over the metal and the nitrogen and carbon of the triple bond:

$$[R-C\equiv N \rightarrow M^{(n+1)+} \equiv R-C\equiv \overset{+}{N}-M^{n+}] \longleftrightarrow R-\overset{+}{C}=N-M^{n+}$$

This co-ordination of the nitrogen not only causes an increase of 50–110 cm^{-1} in the infrared active carbon–nitrogen stretching vibration, but also activates the cyanide group towards the addition of nucleophilic reagents. Indeed, one of the first reported reactions of a co-ordinated organic ligand

(1908) was the addition of water to the triple bonds of the acetonitrile ligands in the complex *cis*-PtCl$_2$(MeCN)$_2$. This addition takes place under the action of aqueous silver nitrate, and is accompanied by oxidation of the metal from platinum(II) to platinum (IV). The actual structure of the resultant deep blue complex ('platinblau') is still unknown, but is thought to be (III). Although the effect of ammonia on *cis*-PtCl$_2$(MeCN)$_2$ is not quite so complicated, as well as adding to the two triple bonds the ammonia also displaces the co-ordinated chloride anions to give the salt (IV).

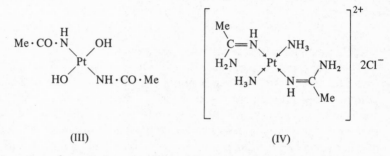

(III)   (IV)

The reaction of ammonia and primary aromatic amines with organic cyanides to give amidines and *N*-substituted amidines respectively is markedly accelerated by Lewis acids such as the chlorides of Al(III), Fe(III) and Sn(IV), which can form complexes with the nitriles.

$$R-C{\equiv}N + Ar \cdot NH_2 \xrightarrow{\text{AlCl}_3} R-C{=}NH$$
$$\underset{Ar \cdot NH}{|}$$

Not unexpectedly, the co-ordinated acetonitrile in the rhenium(IV) complex, ReCl$_4$(MeCN)$_2$, reacts very rapidly with alcohols and aromatic primary amines. If the complex is warmed in ethanol for 30 seconds an excellent yield of the ethyl acetamidate complex (V) is obtained from which the iminoester can be displaced by triphenylphosphine.

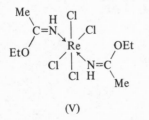

(V)

Similarly, the co-ordinated nitriles in complexes of the type (VI; M = Co, Ru, or Rh) are attacked almost instantaneously by HO$^-$(pH~11) to give the amido complexes (VII). These are reversibly protonated to give the corresponding amide complexes.

$$[(NH_3)_5M \leftarrow N \equiv C \cdot R]^{3+} \xrightarrow{HO^-} [(NH_3)_5M-NH \cdot CO \cdot R]^{2+} \xrightleftharpoons{H^+} [(NH_3)_5M \leftarrow NH_2 \cdot CO \cdot R]^{3+}$$

(VI)   (VII)   (VIII)

Kinetic studies show that the attack by HO^- proceeds at about 10^6 (Co) to 10^8 (Ru) faster than with the unco-ordinated nitrile. With the inexpensive cobalt(III) system, the co-ordinated amide in (VIII) can easily be removed from the metal by reducing the latter with V^{2+} or Cr^{2+}, thus providing a route for converting nitriles into amides that does not require the hot, strong acids or bases normally used.

The co-ordinated nitriles in complexes of type (VI) are also very susceptible to attack by hydride anion, and can be reduced to the primary amine stage, $[(NH_3)_5M(H_2N \cdot CH_2R)]^{3+}$, by sodium borohydride. This allows the selective reduction of cyano groups in compounds such as $NC \cdot CH_2 \cdot CO_2Et$ and $NC \cdot (CH_2)_3 \cdot NO_2$.

In all the metal-promoted additions described above, the nitrogen of the nitrile group is co-ordinated with a metal ion before nucleophilic attack occurred. This is not always the situation, as illustrated in the reactions below.

The most convenient method for converting 2-cyanopyridine into 2-ethoxy-carbonylpyridine is to treat the cyanide with one equivalent of copper(II) chloride in ethanol, and then decompose the resultant complex (X) with hydrochloric acid or aqueous EDTA. The overall yield is good (> 60 per cent), and the formation of the complex (X) is very rapid at room temperature. Cyanide groups in the 3- and 4-positions are not affected, thus allowing 2,4-dicyanopyridine to be readily converted into 4-cyano-2-ethoxycarbonyl-pyridine.

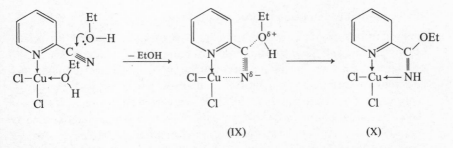

(IX)   (X)

When one considers that the stereochemistry of 2-cyanopyridine prevents both of the nitrogen atoms in this compound from co-ordinating with the same metal ion, the rapidity of the addition of ethanol to the nitrile group initially seems rather surprising. However, the rapidity has been rationalised by assuming that in the transition state of the addition the partly negatively-charged nitrogen of the developing imino anion is quite strongly attached to the metal and has displaced a co-ordinated ethanol molecule (see(IX)). This displacement of a solvent molecule during the reaction and the associated increase in entropy are very important, for they result in this type of addition being thermodynamically more favourable than that in which the nitrile group is already co-ordinated with the metal in the starting complex, and in which a co-ordinated solvent molecule has already been displaced before the addition commences.

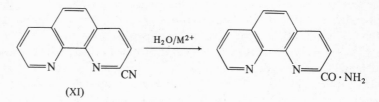

(XI)

This explanation is supported by a kinetic study of the hydration promoted by metal ions of 2-cyanophenanthroline (XI) which is structurally analogous to 2-cyanopyridine. Co-ordination of the phenanthroline with a transition-metal ion considerably increases the rate of alkaline hydrolysis of the cyanide group to the amide. With Ni^{2+}, for example, the second-order rate constant is 10^7 greater than that of the hydration in the absence of the metal, and a comparison of the activation parameters of these two reactions reveals that the increase is entirely due to the difference in the entropies of activation of the two reactions.

Chemical reactions and properties of transition-metal complexes of organic nitriles have been reviewed by Walton (1965) and by Storhoff and Lewis (1977).

4.3. Additions to carbon–carbon triple bonds

Ionic additions to carbon–carbon triple bonds are usually initiated by nucleophilic reagents, and these additions proceed more readily with conjugated poly-yne systems. In contrast, the less-commonly observed electrophilic additions are inhibited by conjugation, and are only very rarely observed with systems that contain more than three conjugated triple bonds. Most of the nucleophilic additions are strongly catalysed by certain metal ions such as Ag^+, Cu^+, Pd^{2+}, Hg^{2+} and Tl^{3+}, and thus the formation of indoles, benzofurans and thionaphthenes from o-substituted arylalkynes in which the substituent is a nucleophilic group (NuH; Nu = NR', O, or S) is carried out in the presence of a copper(I) or silver(I) salt.

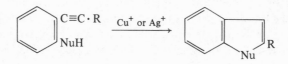

This is also true of the formation of thiophens, selenophens and *N*-substituted pyrroles (XII; Nu = S, Se, or NR respectively) from conjugated diynes.

$$R \cdot C \equiv C \cdot C \equiv C \cdot R' + H_2 Nu \xrightarrow{\text{Cu}^+ \text{ or Ag}^+}$$

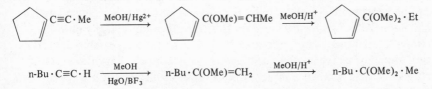

(XII)

The addition of water, alcohols, phenols, and carboxylic acids to a triple bond is often promoted by a mercury(II) compound.

$$Et \cdot C \equiv C \cdot Et \xrightarrow{\text{Hg(OAc)}_2/\text{aq. AcOH}/\text{H}_2\text{SO}_4} n\text{-Pr} \cdot CO \cdot Et$$

$$H \cdot C \equiv C \cdot H + HO \cdot CO \cdot CH_2 Cl \xrightarrow{\text{HgO}} H_2 C = CH \cdot O \cdot CO \cdot CH_2 Cl$$

In these additions, boron trifluoride or a mineral acid is usually also included in the reaction mixture, and a mixture of boron trifluoride, mercury(II) oxide and trichloroacetic acid, and a solution of 5 per cent (w/v) mercury(II) sulphate in 10 per cent sulphuric acid is recommended for the addition of alcohols and water respectively. With alcohols the initial addition is followed by an acid-catalysed addition and ketals are formed.

$$\boxed{} C \equiv C \cdot Me \xrightarrow{\text{MeOH}/\text{Hg}^{2+}} \boxed{} C(OMe) = CHMe \xrightarrow{\text{MeOH}/\text{H}^+} \boxed{} C(OMe)_2 \cdot Et$$

$$n\text{-Bu} \cdot C \equiv C \cdot H \xrightarrow[\text{HgO}/\text{BF}_3]{\text{MeOH}} n\text{-Bu} \cdot C(OMe) = CH_2 \xrightarrow{\text{MeOH}/\text{H}^+} n\text{-Bu} \cdot C(OMe)_2 \cdot Me$$

The mercury(II) catalysed addition of water to acetylene is one of the industrial routes to acetaldehyde.

$$H \cdot C \equiv C \cdot H + H_2 O \xrightarrow{\text{HgSO}_4/\text{aq. H}_2\text{SO}_4} Me \cdot CHO$$

Mercury(II) acetate in acetic acid is often used for converting alkynes into enol acetates which are then hydrolysed to the parent carbonyl compound. In this process, and in the mercury(II)-promoted additions carried out under neutral conditions, the initial addition product is a σ-bonded mercury derivative, the mercury–carbon bond of which can be conveniently cleaved by hydrogen sulphide or hydrogen chloride.

$$Ph \cdot C \equiv C \cdot Ph \xrightarrow{\text{Hg(OAc)}_2/\text{AcOH}} Ph \cdot C(OAc) = C(HgOAc)Ph$$

$$\downarrow H_2 S$$

$$Ph \cdot CO \cdot CH_2 Ph \xleftarrow{\text{H}_3\text{O}^+} Ph \cdot C(OAc) = CH \cdot Ph$$

σ-Bonded derivatives are presumably also formed – and subsequently suffer cleavage *in situ* – during the additions carried out under acidic conditions.

$$R-C\equiv C-R' \xrightarrow{\text{Nu}:/\text{Hg}^{2+}} \underset{R}{\overset{\text{Nu}}{\diagdown}}C=C\underset{\text{Hg}^+}{\overset{R'}{\diagup}} \xrightarrow{\text{H}^+} \underset{R}{\overset{\text{Nu}}{\diagdown}}C=C\underset{\text{H}}{\overset{R'}{\diagup}} + \text{Hg}^{2+}$$

Regardless of whether the initial derivative is actually isolated or not, the products of additions to unsymmetrical alkynes are consistent with the mercury having become attached to the more electronegative carbon of the triple bond, as in the following example.

$$\text{Et} \cdot \text{C} \equiv \text{CH} \xrightarrow{\text{Hg(OAc)}_2/\text{AcOH}} \text{Et} \cdot \text{C(OAc)} = \text{CH(HgOAc)}$$
$$\downarrow \text{H}^+$$
$$\text{Et} \cdot \text{C(OAc)} = \text{CH}_2$$

The synthetic applications of mercury(II)-promoted additions and the various experimental procedures that may be employed have been discussed by Raphael (1955), Winterfeldt (1969), and Pizey (1974).

All of the metal-promoted nucleophilic additions outlined above may be rationalised by assuming that attack by the nucleophile on the carbon-carbon triple bond is facilitated by the triple bond being co-ordinated with a metal ion. Although the evidence for this co-ordination is largely more circumstantial than direct, in one case, i.e. the mercury(II)-catalysed hydration of phenylacetylene, the reaction has been definitely shown to proceed *via* a 2 : 1 alkyne–metal complex. Unlike nitriles which nearly always co-ordinate with metals by σ-bond formation, alkynes co-ordinate by using

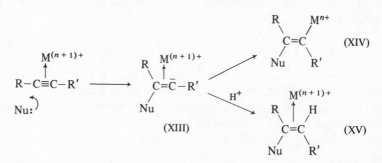

their π-orbitals, and it is reasonable to assume that formation of a metal complex would facilitate nucleophilic attack, for in the resultant carbanion (XIII) the negative charge would be situated on a carbon whose p-orbital was directly involved in the π-bonding. It is to be expected that unless steric effects intervened this carbon would be the more electronegative of the two alkyne carbons, and this is consistent of course with the structures of the final organic compounds obtained from unsymmetrical alkynes.

Presumably in the mercury(II) systems the carbanion (XIII) collapses to give the σ-bonded derivative (XIV), but with metal ions that form rather weak σ-bonds with carbon, e.g. Ag^+, the addition of a proton to give the π-bonded alkene (XV) occurs preferentially.

Most of the additions to alkynes are carried out in acid solution to promote the ionisation of the metal salt that is used, for most of the salts are covalent in nature. With additions of water, alcohols and other weak acids, the acidic conditions probably result in the co-ordinated alkyne being attacked by the undissociated nucleophile, rather than by its conjugate base. Exceptions to this could arise if nucleophilic attack is intramolecular, i.e. when the co-ordinated alkyne reacts with an adjacent negatively-changed ligand, but so far results consistent with this type of mechanism have only been obtained in the work described below.

Of the seven rhodium(III) complexes of composition $[Rh(H_2O)_n Cl_{6-n}]^{(n-3)+}$ the cationic species ($n = 4, 5$ and 6) do not catalyse the hydration of alkynes, and are inert towards substitution of the chloride anions by other ligands. The hexachloro species ($n = 0$) is also inert as a catalyst but is reactive towards substitution. The remaining three species ($n = 1, 2$ and 3) are active as catalysts with the activity a maximum at $n = 1$, and all three undergo substitutions. These observations and the kinetics of the catalysed hydration are consistent with the hydrations being initiated by displacement of a chloride anion by the alkyne. Deprotonation of an adjacent co-ordinated water molecule can occur to some extent even in acid solution because of the increase in acidity caused by co-ordination (see p. 000), and the resultant co-ordinated hydroxide anion can then attack the alkyne intramolecularly.

The complexes $[Ru(H_2O)_2Cl_4]^-$ and $[Ru(H_2O)Cl_5]^{2-}$ are probably involved in an identical manner in the catalysis of the hydration of alkynes by ruthenium(III) chloride in hydrochloric acid, for the rates of hydration are proportional to the sum of the concentrations of these two complexes.

4.4.   Additions to carbon–carbon double bonds

Many additions to alkenes proceed by initial formation of an alkene–electrophile adduct which is then attacked by a nucleophile. The addition of bromine
and the acid-catalysed addition of water are well-established reactions which
proceed by this route, and in these cases the electrophiles that are thought to
be involved are a polarised bromine molecule and a proton respectively.
Certain metal cations can also function as the electrophile and promote
nucleophilic attack on the double bond of an alkene by forming a π-complex.

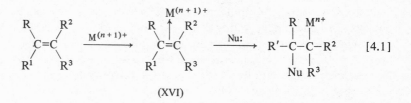

(XVI)

With some systems the π-complex (XVI) is a stable entity which can be
isolated and even purified before it is subjected to nucleophilic attack, but
with others the complex is very labile and often there is only circumstantial
evidence for its participation.

As shown in equation 4.1 the overall product of metal-promoted addition
is a metal alkyl, and accordingly the metal cations which function in this way
are those whose σ-bonds to saturated carbon are relatively stable from the
thermodynamic viewpoint, e.g. Hg^{2+}, Pt^{2+} and Tl^{3+}. Consequently, although
many alkenes form π-complexes with Ag^+, these complexes do not react with
nucleophiles on account of the very low stability of the metal–silver(I) σ-bond.

With the notable exception of mercury(II) alkyls which are discussed in a
separate section (§4.5) on account of their wide application to synthetic
organic chemistry, most of the metal alkyls formed in metal-promoted additions to simple alkenes are chemically unstable, and tend to decompose *in
situ* to give metal-free organic products (see §4.6). Their stability is increased,
however, if the metal is chelated, i.e. if the double bond in the starting
alkene forms part of a chelating ligand. It is metal-promoted additions to this
type of alkene that are discussed in this section.

Not surprisingly, chelating alkenes tend to form relatively stable π-com-
plexes with metal cations, and nucleophilic attack on a stable, isolable π-
complex is well illustrated by reactions of the palladium(II) chelate (XVII)
of cyclo-octa-1,5-diene. This chelate is a stable orange solid, m.p. 210 °C,
and is conveniently prepared by treating the diene with sodium tetrachloropal-
ladate(II), $Na_2(PdCl_4)$. The corresponding platinum(II) compound can be
prepared similarly and, as is usually the case, it is thermally more stable than
the palladium(II) chelate, and its double bonds are less reactive towards
nucleophiles.

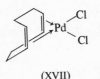

(XVII)

If the palladium chelate is stirred with a suspension of sodium carbonate in methanol at 25 °C, attack by methanol (or methoxide anion) on one of the co-ordinated double bonds causes formation of the organopalladium compound (XVIII) and displacement of a chloride anion. In the compound (XVIII) the fourth co-ordination position of the palladium is filled by the compound existing as a halogen-bridged dimer, which for convenience is

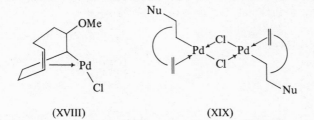

(XVIII) (XIX)

represented by the general structure (XIX; Nu = OMe). This reaction with methanol is reversible, and if the dimer is heated in aqueous hydrochloric acid, the diene complex (XVII) is re-formed.

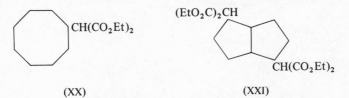

(XX) (XXI)

Other nucleophilic reagents that attack the complex (XVII) and give dimeric products of type (XIX) include sodium azide, ammonia, primary and secondary amines, and hydroxide anion. The dimer (XIX; Nu = $CH(CO_2Et)_2$) obtained when the complex is treated with diethyl malonate and sodium carbonate in ether exhibits a number of reactions typical of these organopalladium addition products. Thus, reduction with sodium borohydride gives the saturated ester (XX), for this reagent not only cleaves the carbon–palladium bond but also reduces the co-ordinated double bond, presumably through the formation of palladium hydride species. Removal of the metal without reduction of the double bond can be achieved by heating the dimer with sodium ethoxide in ethanol, when an excellent yield of diethyl cyclo-oct-4-enylmalonate is obtained. This reductive cleavage of the carbon–palladium bond probably involves an unstable palladium(II) alkoxide.

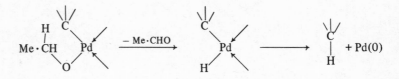

Treatment of the dimer with methylsulphinyl anion (Me . SO . CH_2^-) causes removal of the acidic hydrogen which is flanked by the two ester groups, and the resultant carbanion undergoes an intramolecular nucleophilic substitution at the saturated carbon attached to the metal (see equation 4.2) to give the bicyclo[6.1.0]nonene ester (XXII). A similar nucleophilic substitution, but of a transannular type, occurs during the formation of the bicyclic ester (XXI) from the dimer when the remaining co-ordinated double bonds are attacked by diethyl malonate anion.

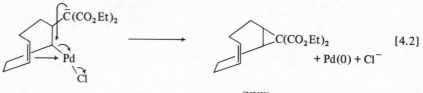

[4.2]

(XXII)

The stereochemistries of the products obtained by the action of most nucleophilic reagents on the complex (XVII) are consistent with rearside attack on the co-ordinated alkene.

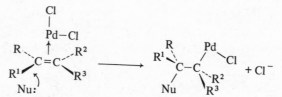

This is also true of the additions which have been reported for the palladium(II) and platinum(II) halide complexes of other chelating alkenes such as norbornadiene and *endo*-dicyclopentadiene. Additions with complexes of the latter alkene involve the more reactive of the two double bonds, and thus dichloro-*endo*-(dicyclopentadiene)platinum(II) reacts with benzylamine to give the complex (XXIII) as a dimer.

The compound (XXIV) obtained (also as a dimer) by the action of methanol on the palladium(II) dichloride complex of norbornadiene is noteworthy in that when treated with 1,2-bis(diphenylphosphino)ethane it gives the σ-bonded complex (XXV). This reacts with lithium aluminium hydride and with bromine to give the nortricyclenes (XXVI; X = H and Br respectively).

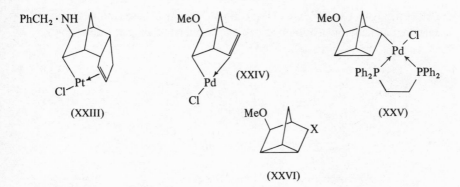

(XXIII)   (XXIV)   (XXV)

(XXVI)

The discussion so far has been confined to reactions of stable complexes that have been isolated and purified before being subjected to nucleophilic attack. Such reactions are in fact atypical, for most reactions in which a co-ordinated double bond is attacked by a nucleophile involve complexes that are formed *in situ*. Some of these complexes are probably fairly stable, particularly those of alkenes in which the double bond is one of the donors of a di- or multi-dentate ligand. The following transformations are of alkenes of this type.

When 2-vinylpyridine reacts with sodium tetrachloropalladate(II) in methanol, the product is a dimer of the σ-bonded complex (XXVII). This on reduction with lithium aluminium hydride gives 2-(1′-methoxyethyl)-

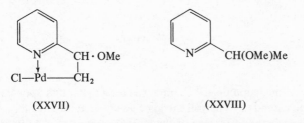

(XXVII)   (XXVIII)

pyridine (XXVIII), and the corresponding ethyl and isopropyl ethers can be prepared in a similar manner. This indirect addition of alcohols to the double bond of vinylpyridine contrasts with the base-catalysed addition, which affords the isomeric 2-(2′-alkoxyethyl)pyridines.

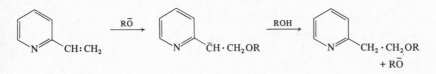

Other σ-bonded complexes that can be formed by the reaction between a chelating alkene, a metal ion and a nucleophile, include (XXIX; Nu = OMe,

$CH(CO_2Et)_2$ or $CH(CO_2Et)$. COMe) and (XXX), and these complexes almost certainly arise by nucleophilic attack on the chelated alkene. Interestingly,

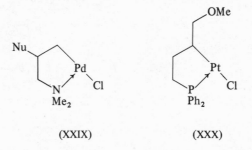

(XXIX)   (XXX)

the complex (XXXII) can be prepared by the action of methoxide anion on the dichloride (XXXI), even though the double bonds in the latter are not co-ordinated. Possibly this reaction proceeds by a route in which the double bonds become co-ordinated, or one in which nucleophilic attack on an unco-ordinated double bond and bonding of the resultant negatively-charged carbon to the metal are concerted.

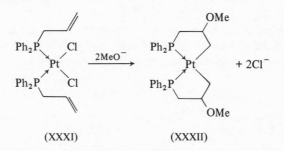

(XXXI)   (XXXII)

An example of nucleophilic attack on a chelating diene that appears capable of being extended to permit the synthesis of a number of interesting terpenoid systems, is the formation of the π-allylic complex (XXXIII) (represented for convenience as the σ-bonded monomer) from myrcene, methanol and sodium tetrachloropalladate(II).

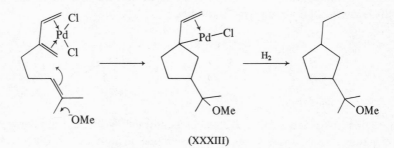

(XXXIII)

4.5. Mercury (II)-promoted additions to alkenes

In the presence of mercury(II) salts, alkenes react with a variety of nucleophiles to give β-substituted mercury alkyls.

$$\begin{array}{c} R \\ \diagdown \\ \diagup \\ R^1 \end{array} C=C \begin{array}{c} R^2 \\ \diagup \\ \diagdown \\ R^3 \end{array} \xrightarrow{\text{Nu:}/\text{Hg}^{2+}} \quad R^1-\overset{\displaystyle R}{\underset{\displaystyle Nu}{C}}-\overset{\displaystyle Hg^+}{\underset{\displaystyle R^3}{C}}-R^2$$

When the substituent is bonded to the β-carbon atom *via* oxygen, e.g. OH, OR or O . CO . R, the process of converting the alkene into the β-substituted mercury alkyl is termed 'oxymercuration'. If one wishes to be more specific, the terms 'hydroxymercuration', 'acetoxymercuration', etc., can be used in order to indicate the nature of the β-substituent. Oxymercuration reactions have been reviewed by Chatt (1951), and more recently by Kitching (1968) and Bloodworth (1977).

It is generally considered that the mercury alkyl is formed from the alkene by nucleophilic attack on a intermediate mercurinium ion, i.e. a π-complex between the alkene and a mercury cation. When mercury(II) acetate in water is used the alkene is probably complexed with a hydrated Hg^{2+} ion, for the acetate is completely ionised in this solvent, but with the highly covalent $HgCl_2$ an alkene/Hg^+Cl complex seems more likely.

$$HgCl_2 \rightleftharpoons Hg^+Cl + Cl^-$$

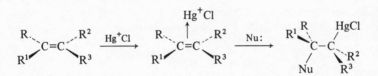

Extended Hückel molecular orbital calculations indicate that unlike the π-complexes formed from alkenes and transition-metal ions, the complexes formed from alkenes and mercury(II) are hardly stabilised at all by backbonding. In the case of the ethylene/Hg^{2+} complex, 97 per cent of the total bonding between the metal and ligand appears to arise from overlap of the alkene π-orbitals with the empty 6s-orbital on the mercury. Because of the absence of back-bonding the co-ordinated alkene bears a substantial positive charge, and it is therefore not surprising that the complex reacts rapidly with nucleophiles.

At this point it must be mentioned that at the present time there is in fact no direct evidence that alkene/mercury(II) complexes are formed in mercuration reactions, and all of the evidence for their participation is largely circumstantial. It is known, however, that such complexes are capable of existence, for they can be generated from alkenes by using specialised conditions such

(XXXIV)   (XXXV)

as highly acidic solvents of low nucleophilicity. The complex (XXXIV), for example, can be generated by adding cyclohexene in liquid sulphur dioxide to a solution of methylmercury(II) fluorosulphate in a fluorosulphonic acid/sulphur dioxide mixture. Apart from the metal–proton coupling which is normally not observed with silver complexes, the ^{1}H n.m.r. spectrum of (XXXIV) is identical with that of (XXXV), while the ^{13}C n.m.r. spectra of the two complexes show that the vinylic carbons in the mercury system are more deshielded than those in the silver one, probably due to the decreased degree of back-bonding from the metal.

The preparation of β-substituted mercury alkyls from alkenes is synthetically useful to the organic chemist because the carbon–mercury bond in these alkyls can be cleaved by halogens and by reducing agents to give organic products in which the mercury has been replaced by a halogen and a hydrogen respectively. Of the various reducing agents that can be used, e.g. sodium amalgam, lithium aluminium hydride, hydrazine, and sodium borohydride, the last is the most useful and does not require the mercury alkyl to be isolated prior to the reduction. In the preparation of alcohols from alkenes, for example, the alkene is treated with mercury(II) acetate in aqueous tetrahydrofuran, and the resultant mercury alkyl is reduced *in situ* by the addition of sodium borohydride in aqueous sodium hydroxide. Excellent yields of alcohols are obtained after total reaction times of less than 30 minutes.

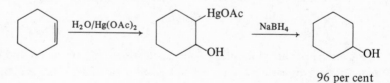

96 per cent

The same reaction sequence but with methanol, ethanol, or isopropanol as the solvent in the first stage affords the appropriate ether. With sterically hindered alcohols such as t-butanol, however, mercury(II) acetate is an unsatisfactory reagent for generating the intermediate π-complex, because the rate at which the complex reacts with the alcohol is comparable with the rate at which it reacts with the acetate anion. Mercury(II) trifluoroacetate is therefore used instead on account of the much lower nucleophilicity of the trifluoroacetate anion. Use of the feebly nucleophilic anion NO_3^- is essential for the π-complex to react successfully with acetonitrile. With this combination, reduction of the resultant mercury alkyl affords an *N*-substituted acetamide which can be hydrolysed to the parent amine.

$$\text{n-C}_4\text{H}_9 \cdot \text{CH} = \text{CH}_2 \xrightarrow{\text{Hg(NO}_3)_2/\text{MeCN}} \underset{\overset{|}{\underset{\overset{|}{\text{O} \cdot \text{NO}_2}}{\text{N} = \text{C} \cdot \text{Me}}}}{\text{n-C}_4\text{H}_9 \cdot \text{CH} - \text{CH}_2 \cdot \text{HgNO}_3}$$

$$\Bigg\downarrow \text{aq. NaBH}_4$$

$$\underset{\overset{|}{\text{NH} \cdot \text{CO} \cdot \text{Me}}}{\text{n-C}_4\text{H}_9 \cdot \text{CH} - \text{CH}_3}$$

Other nucleophiles that can be used in conjunction with mercury(II) acetate include t-butylhydroperoxide.

$$\text{R} \cdot \text{CH} = \text{CH}_2 \xrightarrow[\text{(2) NaBH}_4]{\text{(1) Me}_3\text{C} \cdot \text{O} \cdot \text{OH/Hg(OAc)}_2} \underset{\overset{|}{\text{O} \cdot \text{O} \cdot \text{CMe}_3}}{\text{R} \cdot \text{CH} \cdot \text{CH}_3}$$

In all these mercury-promoted reactions between nucleophiles and non-conjugated alkenes, the nucleophile becomes attached to the more electropositive carbon of the double bond, i.e. when the resultant mercury alkyl is reduced the structure of the product is in accord with Markownikov's rule. The oxymercuration process therefore provides a convenient alternative to the acid-catalysed addition of water and alcohols for converting alkenes into alcohols and ethers respectively. The hydroxymercuration process has the additional advantage that as carbonium ions are not involved, rearrangements are not encountered.

$$\text{Me}_3\text{C} \cdot \text{CH} = \text{CH}_2 \xrightarrow[\text{(2) NaBH}_4]{\text{(1) MeOH/Hg(O} \cdot \text{CO} \cdot \text{CF}_3)_2} \text{Me}_3\text{C} \cdot \text{CH(OMe)} - \text{CH}_3$$

When $\alpha\beta$-unsaturated carbonyl compounds are submitted to oxymercuration reactions, the nucleophile preferentially attacks the β-carbon atom except when only the α-carbon atom of the double bond is substituted and bears an alkyl substituent.

$$\text{Me} \cdot \text{CH} = \text{CH} \cdot \text{CO}_2\text{Me} \xrightarrow{\text{MeOH/Hg(OAc)}_2} \text{Me} \cdot \text{CH(OMe)} \cdot \text{CH(HgOAc)} \cdot \text{CO}_2\text{Me}$$

$$\text{H}_2\text{C} = \text{CMe} \cdot \text{CO}_2\text{Me} \xrightarrow{\text{MeOH/Hg(OAc)}_2} \text{H}_2\text{C(HgOAc)} \cdot \text{C(OMe)Me} \cdot \text{CO}_2\text{Me}$$

With unsymmetrical alkenes in which there is very little or no difference between the electronegativities of the two carbons of the double bond, an approximately equal mixture of the two possible isomers is formed. Thus the product obtained by methoxymercuration of methyl oleate followed by reduction of the mercury alkyls by borohydride is a 1 : 1 mixture of methyl 9- and 10-methoxystearates. However, the proportions of the two isomers formed in such an oxymercuration are often affected by adjacent substituents which, although not sufficiently close to the double bond to alter the electronegativities of the two unsaturated carbon atoms can preferentially direct the attack of the nucleophile by co-ordinating with the mercury, either in the π-complex itself (as shown in (XXXVI)) or in the transition state. Only weak co-ordination is required to cause substantial deviation from a 1 : 1 mixture of the two possible products.

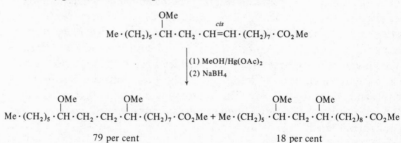

(XXXVI)

Substituents that can act in this way include $-OH$, $-OR$ and CO_2R, in the β- or γ-position with respect to the double bond.

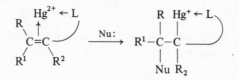

$$Me \cdot (CH_2)_5 \cdot \overset{OMe}{\underset{|}{CH}} \cdot CH_2 \cdot \overset{cis}{CH=CH} \cdot (CH_2)_7 \cdot CO_2 Me$$

(1) MeOH/Hg(OAc)$_2$
(2) NaBH$_4$

$$Me \cdot (CH_2)_5 \cdot \overset{OMe}{\underset{|}{CH}} \cdot CH_2 \cdot CH_2 \cdot \overset{OMe}{\underset{|}{CH}} \cdot (CH_2)_7 \cdot CO_2 Me + Me \cdot (CH_2)_5 \cdot \overset{OMe}{\underset{|}{CH}} \cdot CH_2 \cdot \overset{OMe}{\underset{|}{CH}} \cdot (CH_2)_8 \cdot CO_2 Me$$

79 per cent   18 per cent

In the mercuration of most alkenes the stereochemistry of the final organic product indicates an overall *trans*-addition to the double bond. This is well illustrated by cyclohexene systems, where almost entirely *trans*-diaxial addition is observed.

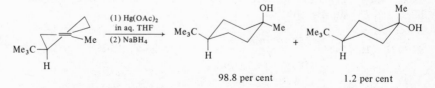

98.8 per cent   1.2 per cent

As reduction of the intermediate mercury alkyl by borohydride is known to occur with retention of configuration at carbon, this stereochemistry is consistent with the co-ordinated double bond in the intermediate π-complex being attacked by the nucleophile on the opposite side to the metal, i.e. as in other systems (see p. 203).

Exceptions to *trans*-addition arise in the reactions with alkenes in which the rearside of the co-ordinated double bond is sterically hindered, e.g. *trans*-cyclo-octene, and with alkenes in which the double bond is highly strained. Norbornene, for example, gives a 100 per cent yield of the *cis*-exo product.

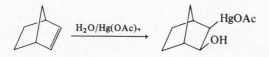

This stereochemistry of the mercuration of strained double bonds is also characteristic of the addition of other electrophilic reagents such as hydrogen chloride. The possible reasons for this anomaly have been discussed by Traylor (1969).

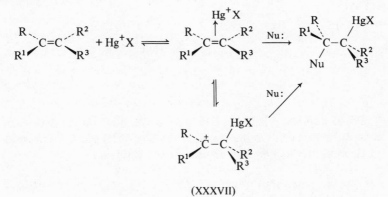

(XXXVII)

The absence of rearrangement products when highly alkylated alkenes and strained alkenes are submitted to oxymercuration reactions clearly shows that any intermediate π-complex is not converted into a carbonium ion of type (XXXVII). This appears to be true even when the cation (XXXVII) would be resonance-stabilised, for the methoxymercuration of the optically active allene (XXXVIII) is a highly stereospecific reaction, and gives only the two diastereomers (XXXIX) and (XL), thus eliminating the possibility that the carbonium ion (XLI) is the reaction intermediate.

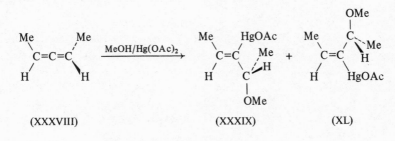

(XXXVIII)   (XXXIX)   (XL)

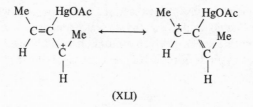

(XLI)

All stages of the oxymercuration reactions are reversible, and the mercury alkyls produced can be converted back into the starting alkene by treatment with mineral acids or pyridine, i.e. reagents that can supply, or that are themselves, strong donor ligands for the Hg^{2+} ion produced from the reverse reaction. The reversibility of the oxymercurations allows a number of nucleophilic substitutions at a vinylic carbon to be catalysed by mercury(II) salts. The reaction sequence is analogous to that involved when palladium(II) salts are used (see p. 208).

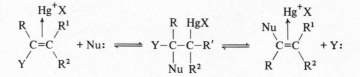

For example, vinyl acetate is converted into other vinyl esters when heated with carboxylic acids in the presence of mercury(II) acetate, and vinyl ethers of alcohols may be prepared by use of ethyl vinyl ether.

$$H_2C{=}CH{\cdot}O{\cdot}CO{\cdot}Me + R{\cdot}CO_2H \xrightarrow{\text{Hg(OAc)}_2} H_2C{=}CH{\cdot}O{\cdot}CO{\cdot}R + Me{\cdot}CO_2H$$

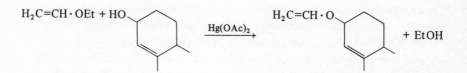

The hydrolysis of vinyl sulphides to aldehydes or ketones is similarly catalysed.

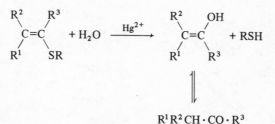

$$R^1R^2CH{\cdot}CO{\cdot}R^3$$

Normally, fairly drastic conditions are necessary to carry out this hydrolysis, but if mercury(II) chloride in refluxing aqueous acetonitrile is used, high yields of the carbonyl compound are obtained after comparatively short reaction times.

The ease with which a β-alkoxy- or carboxy-mercury alkyl is converted back into an alkene also allows a convenient two-step sequence for regenerating carboxylic acids from their cinnamyl esters.

$$\underset{\underset{\text{CH}_2 \cdot \text{CH=CHPh}}{|}}{\text{R} \cdot \text{CO} \cdot \text{O}} \quad \xrightarrow{\text{MeOH/Hg(OAc)}_2} \quad \underset{\underset{\underset{\text{HgOAc}}{|}}{\underset{\text{CH}_2-\text{CH}-\text{CHPh}}{|}}}{\text{R} \cdot \text{CO} \cdot \text{O} \qquad\quad \text{OMe}}$$

$$\downarrow \text{NCS}^-$$

$$\text{R} \cdot \text{CO}_2\text{H} + \text{H}_2\text{C=CH} \cdot \text{CH(OMe)Ph}$$

Accordingly, the cinnamyl group has been proposed as a protecting group for carboxylic acids.

4.6.   Unstable metal alkyls produced by nucleophilic attack on co-ordinated alkenes

As mentioned on p. 190 nucleophilic attack on the π-complex formed between a metal cation and a simple alkene often affords an unstable metal alkyl that decomposes *in situ* to give metal-free organic products. The most frequent mechanisms by which this decomposition occurs are shown in figure 4.1, p. 202, and it can be seen that all of them involve heterolytic fission of the metal–carbon σ-bond, i.e. the metal functions as a leaving group. Indeed, in every case the mechanisms correspond to those of analogous 'organic' reactions in which the group (M) is a conventional leaving group such as a halide or tosylate anion.

A further point is that for clarity all of the mechanisms are represented as two-stage processes. In the first stage heterolytic fission of the metal–carbon bond affords a carbonium ion, and in the second stage this carbonium ion either loses a proton, suffers nucleophilic attack, or undergoes a 1,2-shift of a β-substituent. In practice, of course, both steps may be concerted, as illustrated in some of the examples described below.

The overall reaction between a nucleophile and a co-ordinated alkene to give a metal alkyl which then decomposes by one of the mechanisms shown in

$$\underset{\underset{\text{Nu:}}{\uparrow}}{\text{R}^1\text{R}^2\text{C=CR}^3\text{R}^4} \overset{\text{M}^{(n+2)+}}{\longrightarrow} \underset{\underset{\text{Nu}}{|}}{\text{R}^1\text{R}^2\text{C}-\text{CR}^3\text{R}^4} \overset{\text{M}^{(n+1)+}}{\longrightarrow} \text{M}^{n+} + \text{organic products} \quad [4.3]$$

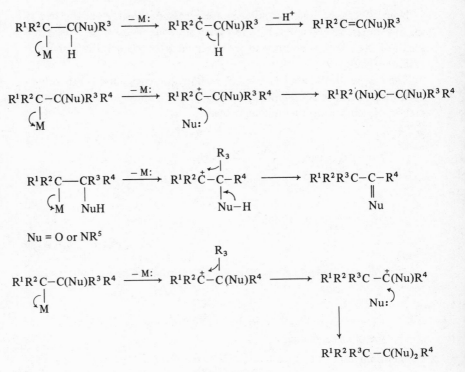

Nu = O or NR⁵

Figure 4.1.

figure 4.1 results in reduction of the metal and oxidation of the alkene (see equation 4.3). This is one route, therefore, by which alkenes are oxidised by metal cations, and in fact is the route by which ethylene is oxidised to acetaldehyde on a commercial scale by palladium(II) salts:

$$H_2C=CH_2 + H_2O + PdCl_2 \rightarrow Me \cdot CHO + 2HCl + Pd(0)$$

This reaction can be made pseudo-catalytic by using a small quantity of the palladium salt together with an oxidising agent that will convert the palladium(0) back into palladium(II). Copper(II) chloride is preferred for this, because after it has oxidised the palladium it can be regenerated by passing air through the reaction mixture:

$$Pd(0) + 2 CuCl_2 \rightarrow PdCl_2 + 2 CuCl$$

$$2 CuCl + 2 HCl + \tfrac{1}{2} O_2 \rightarrow 2 CuCl_2 + H_2O$$

The overall reaction is then simply the formation of acetaldehyde from ethylene and oxygen:

$$2 H_2C = CH_2 + O_2 \rightarrow 2 Me \cdot CHO$$

This oxidation sequence was developed by the German chemical company Wacker-Chemie, and is consequently now known as the Wacker process. Commercially the oxidation is carried out in either one or two stages. In the one-stage process air and ethylene are passed together into an aqueous solution of copper(II) and palladium(II) chlorides and the acetaldehyde is subsequently isolated, while in the two-stage process the ethylene is oxidised in the absence of air and the acetaldehyde is isolated before air is passed in to regenerate the copper(II) salt.

On the basis of kinetic and labelling studies (see Maitlis, 1971, for useful review), it has been deduced that the acetaldehyde is formed by decomposition of an intermediate β-hydroxyethylpalladium(II) species (XLII; other ligands are omitted from the metal) generated by nucleophilic attack by water on co-ordinated ethylene – which is probably in the form of the complex $PdCl_2(H_2O)(C_2H_4)$. The acetaldehyde is always accompanied by 2-chloroethanol, and this is also formed from the intermediate (XLII), this time by nucleophilic attack by chloride anion – a process known to result in inversion of configuration at the saturated carbon involved. Use of *trans*-$[1,2-^2H]$ethylene in the Wacker process results in the chloroethanol having the *threo* structure (XLIV), thus establishing that in the initial nucleophilic attack by water on co-ordinated ethylene the nucleophile approaches the ethylene from the opposite side to the metal (see (XLIII)). This stereochemistry is therefore directly analogous to that observed with other systems (see pp. 198 and 205).

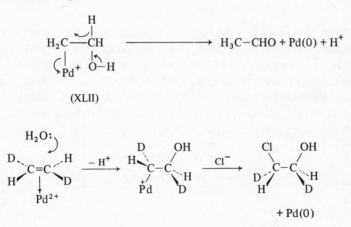

(XLII)

(XLIII)   (XLIV)

The use of palladium(II) salts for oxidising carbon–carbon double bonds has been applied to a number of alkenes other than ethylene (see Bird, 1966). A 1 : 6 mixture of water and dimethylformamide has been found to be the most suitable solvent for alkenes whose solubility in water is very low.

With simple mono-substituted alkenes the oxidation product is predominantly the methyl ketone, in accord with initial nucleophilic attack by water on the more electropositive carbon of the co-ordinated double bond. Dodec-1-ene, for example, gives a mixture of carbonyl compounds (> 80 per cent yield) of which more than 96 per cent is dodecan-2-one. Di- and tri-substituted alkenes tend to give rather complex mixtures, and oxidation of allylic methyl and methylene groups is frequently encountered:

$$Me_2C=CH \cdot CH_2 \cdot Me \xrightarrow{PdCl_2/aq.\ AcOH} Me_2C=CH \cdot CO \cdot Me + Me_2CH \cdot CO \cdot CH_2Me$$

19 per cent   1 per cent

It seems likely that this oxidation occurs *via* a palladium(II) π-allyl complex, for in some cases the same oxidation products are observed if the π-allyl complex of the alkene is first prepared and then oxidised with palladium(II) chloride.

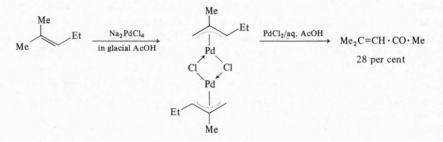

The possible formation of π-allyl complexes also explains (see p. 261) the isomerisation which occurs during the oxidation of some alkenes.

Other transition-metal ions that are effective at oxidising alkenes to saturated carbonyl compounds include Pt^{4+}, Rh^{3+}, Ir^{3+} and Tl^{3+}. With the last ion the carbonyl compounds are sometimes associated with substantial proportions of 1,2-diols which result from one of the alternative routes by which the intermediate metal alkyl can decompose.

$$Me \cdot CH_2 \cdot CH=CH_2 \xrightarrow[\text{in aq. AcOH}]{Tl(OAc)_3} Me \cdot CH_2 \cdot CH(OH) \cdot \overset{\overset{\displaystyle Tl^{2+}}{|}}{CH_2} \longrightarrow Me \cdot CH_2 \cdot CO \cdot CH_3$$

75 per cent

$$Me \cdot CH_2 \cdot CH(OH) \cdot CH_2OH$$

16 per cent

Other nucleophiles that can attack co-ordinated alkenes to give unstable metal alkyls include ammonia and amines. When the metal involved is platinum(II), however, the metal alkyls are often sufficiently stable to permit isolation, and this has enabled the stereochemistry of nucleophilic attack on the co-ordinated alkene to be determined.

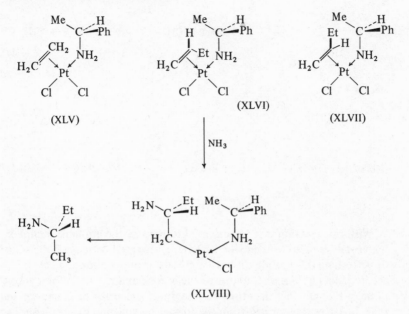

(XLV)　　　　　　(XLVI)　　　　　　(XLVII)

(XLVIII)

Displacement by 1-butene of the co-ordinated ethylene in the complex (XLV) affords a mixture of the two diastereomers (XLVI) and (XLVII) from which the former can be obtained pure by repeated crystallisation. The structure of this complex has been unambiguously determined by X-ray diffraction. The reaction between this diastereomer and ammonia gives a platinum alkyl (probably a dimer of (XLVIII)) which on degradation with concentrated hydrochloric acid gives sec-butylamine with the S-configuration. A similar reaction between the complex (XLVI) and diethylamine gives (S)-*N,N*-diethylbutylamine. The S-configuration of these two amines clearly shows that in both reactions nucleophilic attack on the co-ordinated alkene takes place from the side opposite to the metal.

The σ-bonded alkyl derivatives formed by nucleophilic attack of primary amines on alkenes co-ordinated to palladium(II) are less stable than those formed in the platinum(II) systems described above, and they decompose in a manner similar to that by which the hydroxyethyl derivative (XLII) (p. 203) gives acetaldehyde. For example, the reaction between n-butylamine and the ethylene/palladium(II) chloride complex $[PdCl_2(C_2H_4)]_2$ (XLIX) gives the Schiff base (LI) *via* a β-aminoethyl derivative of type (L).

The use of alcohols as nucleophiles to attack co-ordinated alkenes ultimately affords acetals, as in the reaction between the ethylene complex (XLIX) and primary alcohols.

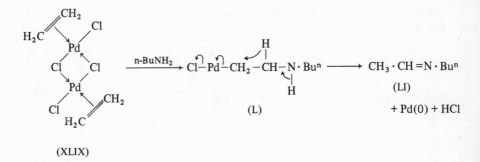

With this reaction it is not essential to prepare the ethylene complex, for the acetals are also formed when ethylene is treated with alcoholic solutions of palladium(II) chloride. In both methods of preparation the acetal is accompanied by minor amounts of the corresponding vinyl ether, and it was initially thought that the ether was the direct precursor of the acetal, and that the latter was formed from the former by addition of the alcohol across the double bond. This idea has definitely been discounted in the formation of the dimethyl acetal, for if the preparation is carried out in MeOD no deuterium is incorporated into the acetal. Presumably the vinyl ether is formed from the intermediate β-alkoxyethyl derivative by a β-elimination.

$$\text{Cl}-\text{Pd}-\text{CH}_2-\overset{\overset{\text{H}}{|}}{\text{CH(OR)}} \longrightarrow \text{CH}_2=\text{CH}\cdot\text{OR} + \text{Pd(0)} + \text{HCl} \qquad [4.5]$$

Substituted alkenes are similarly converted into acetals or ketals by the action of alcoholic solutions of palladium(II) chloride, while use of vicinal diols as the solvent gives the expected 1,3-dioxolane.

$$\text{NC}\cdot\text{CH}=\text{CH}_2 \xrightarrow{\text{MeOH/PdCl}_2} \text{NC}\cdot\text{CH}_2\cdot\text{CH(OMe)}_2$$

$$n\text{-C}_4\text{H}_9\text{CH}=\text{CH}_2 + \underset{\text{OH OH OH}}{\boxed{}} \xrightarrow{\text{PdCl}_2} n\text{-C}_4\text{H}_9-\underset{\underset{\underset{\text{CH}_2\text{OH}}{\big|}}{\text{O}\quad\text{O}}}{\overset{|}{\text{C}}}-\text{Me}$$

As in the formation of acetaldehyde from ethylene, all of these reactions can be made pseudo-catalytic by incorporating into the reaction mixture an oxidant such as copper(II) chloride which can oxidise the palladium(0) back to palladium(II).

Thallium(III) salts are particularly effective at converting alkenes into acetals or ketals, and in the case of thallium(III) nitrate the reactions in methanol are generally complete within a few minutes and give excellent yields.

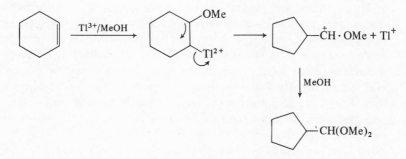

Another nucleophile that can be used to generate unstable metal alkyls from co-ordinated alkenes is the carboxylate anion, and thus while ethylene does not react with palladium(II) chloride in acetic acid the addition of acetate anion results in the formation of metallic palladium, vinyl acetate, ethylidene diacetate, acetaldehyde and acetic anhydride from the unstable β-acetoxyethyl derivative (LII). As this reaction can also be made pseudo-catalytic by adding suitable oxidising agents, e.g. copper(II), and as the vinyl acetate becomes the major product when the reaction is carried out under pressure, this reaction has been developed commercially into a continuous process for preparing the vinyl ester.

$$Cl-Pd-CH_2-CH_2OAc \xrightarrow{\;-H^+\;} CH_2=CH \cdot OAc + Pd(0) + Cl^-$$

(LII) $\Big\downarrow AcO^-$

$$CH_3 \cdot CH(OAc)_2 + Pd(0) + Cl^-$$

Presumably the vinyl ester and the ethylidene diacetate are formed from the β-acetoxyethyl derivative (LII) by the same routes that afford vinyl ethers and acetals when alcohols are used as the nucleophile (see equations 4.5 and 4.4 respectively), while both the acetaldehyde and the acetic anhydride could result from further nucleophilic attack by acetate anion.

$$Cl-Pd-CH_2-\underset{|}{\overset{H}{CH}}-O-\underset{\overset{\|}{O}}{C}-CH_3 \;(AcO^-) \longrightarrow CH_3 \cdot CHO + (CH_3 \cdot CO)_2O + Pd(0) + Cl^-$$

Ethylene can be converted into other vinyl esters by using the appropriate carboxylate anion to attack the palladium(II)/ethylene complex initially formed in solution. Vinyl hexanoate, crotonate and benzoate are examples of esters which have been prepared. Higher alkenes have been subjected to the same type of reaction, mainly with acetic acid as a co-solvent and acetate anion as the nucleophile, but the products formed have been found to be highly dependent upon the alkene, the solvent composition, the reaction temperature, and the inorganic salts which have often been added (see Bird (1966) and Stern (1968)). When copper(II) chloride is added in order to oxidise the palladium(0) which is formed back into palladium(II), chloro-acetates are often major products. The most convenient method currently available for preparing *syn*-7-acetoxynorbornene (LV) is by dehydrochlorina-tion of the 2-chloro derivative (LIV) formed in 84 per cent yield when norbornene is treated with palladium(II) chloride and an excess of copper(II) chloride in acetic acid. The 2-chloro derivative probably arises from the carbonium ion formed by decomposition of the initial σ-bonded intermediate (LIII).

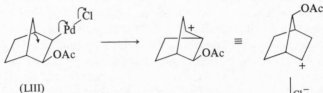

(LIII)

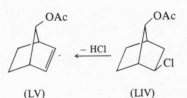

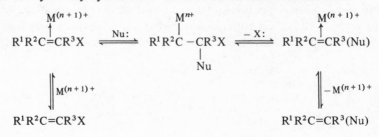

(LV)   (LIV)

Finally it should be noted that with some systems the formation of a σ-bonded organometallic by nucleophilic attack on a co-ordinated alkene is a reversible process, and this allows nucleophilic *substitutions* at a double bond to proceed very readily by successive addition and elimination reactions.

Vinyl acetate is obtained in high yield when vinyl chloride is heated in acetic acid or dimethylformamide with sodium acetate and catalytic amounts of palladium(II) chloride.

$$H_2C=CH \cdot Cl + AcO^- \xrightarrow{\ PdCl_2\ } H_2C=CH \cdot OAc + Cl^-$$

The use of deuteriated vinyl chloride gives results consistent with an addition–elimination mechanism. The same mechanism explains why the 1,2-diacetoxyethylene obtained from 1,2-dichloroethylene in the same manner has the *trans*-configuration regardless of which stereoisomer of the dichloro compound is used, for any *cis*-diacetoxy compound can be converted into the more stable *trans*-form.

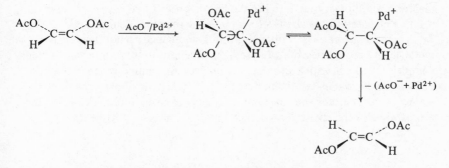

Addition followed by elimination also accounts for the strong catalysis by palladium(II) chloride of the displacement of chloride anion from allylic chlorides by acetate anion.

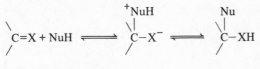

4.7. Reversible formation of imines from primary amines and carbonyl compounds

The addition of nucleophilic reagents such as water, alcohols and amines to carbon–oxygen and carbon–nitrogen double bonds under neutral conditions may be represented by the equation:

$$\begin{array}{ccc} \diagdown & \overset{+}{\underset{|}{N}uH} & Nu \\ C=X + NuH & \rightleftharpoons & \overset{|}{\underset{|}{C}}-X^- \rightleftharpoons & \overset{|}{\underset{|}{C}}-XH \\ \diagup & & \end{array}$$

X = O or NR

Almost all of the kinetic work which has been carried out on the effect of metal ions on these additions has concerned the reversible formation of imines from carbonyl compounds and amines involved in transamination (see p. 101), or in model systems for this process. In the formation of an imine, addition of the amine to the carbon–oxygen double bond of the carbonyl compound is followed by dehydration of the resultant amino alcohol.

$$\backslash C{=}O + H_2N{\cdot}R \rightleftharpoons \overset{\overset{\textstyle NHR}{|}}{\underset{}{\backslash}} C{-}OH \rightleftharpoons \backslash C{=}NR + H_2O \qquad [4.6]$$

Many metal ions displace the equilibrium position of equation 4.6 by forming a complex with the imine. Not all of these metals kinetically catalyse the formation of the imine, and in the formation of *N*-salicylideneglycine from salicylaldehyde and glycine, for example, Cu^{2+} and Ni^{2+} are kinetically inactive even though they stabilise the imine towards hydrolysis by converting it into a chelate of type (LVI). However, some metal ions such as Pb^{2+} and Zn^{2+} that do not possess unfilled d-orbitals, and hence do not have rigid metal–ligand geometries, do catalyse the condensation by forming the mixed complex (LVII), in which although the nucleophilic amino group is co-ordinated with the metal, the bonding is sufficiently flexible to enable the amino group to attack the co-ordinated aldehyde group intramolecularly and ultimately give the chelated imine (LVI).

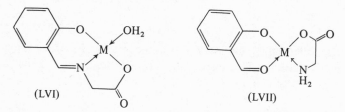

(LVI)   (LVII)

In contrast to the stabilisation of imines mentioned above, several examples are known of imines and carbonyl compounds that when chelated, readily add water and alcohols across the carbon–nitrogen or carbon–oxygen double bond to give stable, chelated addition products. In the presence of halide anions, for example, the cation (LVIII) dissolves in methanol to give (LIX), salts of which do not undergo the reverse elimination of methanol even when heated *in vacuo* at 100 °C.

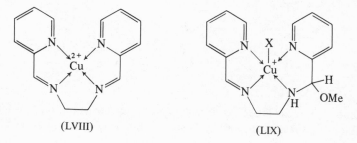

(LVIII)   (LIX)

Analogously, attempts to prepare copper(II) chelates of di-(2-pyridyl) ketone in aqueous media afford salts of the cation (LX) in which the ligand is the hydrated form of the ketone.

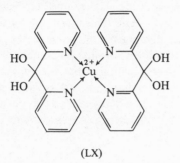

(LX)

The stability of these chelated addition products relative to the chelated unsaturated systems is almost certainly due to the fact that in the latter the ligands involved are planar, and co-ordination with the metal causes substantial angle strain. This strain is relieved to some extent when the addition takes place, i.e. when a trigonal carbon is converted into a tetrahedral one, and the ligand becomes non-planar.

4.8. Asymmetric additions of Grignard reagents and organolithium compounds to carbonyl groups

A reaction between a carbonyl compound and an organometallic reagent derived from a metal whose co-ordination number is higher than its oxidation state must involve complexes at all stages if all the co-ordination sites on the metal are to be permanently filled. In reactions of Grignard reagents the co-ordination number of the magnesium is always maintained at least at four by the metal acting as an acceptor for additional ligands which can be either a solvent molecule, the carbonyl oxygen of the compound undergoing addition, or a halogen atom of another molecule of Grignard reagent or of a molecule of magnesium halide. Examples of these possibilities are illustrated in structures (LXI) and (LXII), the second of which represents the actual complex now thought to be the active species involved in additions of Grignard reagents carried out in ether.

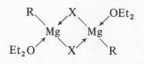

       (LXI)   (LXII)

Because the Mg^{2+} in a solution of Grignard reagent invariably has a co-ordination number of at least four, asymmetric synthesis has often been observed in Grignard reactions when ligands that contain chiral centres have been present, for by co-ordinating with the magnesium these ligands convert the Grignard reagent into a chiral species. This is also true of the analogous reactions with organolithium compounds, for these reagents also accept additional ligands in order to raise the co-ordination number of the Li^+ to four. For example, reactions between achiral carbonyl compounds and Grignard reagents or organolithium compounds are stereoselective to some extent when the chiral ethers (LXIII; R = H or NMe_2) are used as co-solvents or when the reactions are carried out in the presence of (−)-sparteine (LXIV).

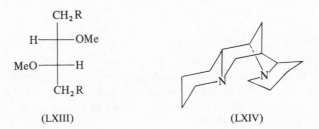

(LXIII)   (LXIV)

In these systems the optical yields are usually fairly low (<10 per cent), but occasionally figures of up to 35 per cent have been reported. Thus, in the formation of 1-phenylpentan-1-ol from benzaldehyde and n-butyl-lithium in a mixture of n-pentane and (−)-sparteine at −130 °C, the (R)-alcohol is the predominant enantiomer with an optical yield of 33 per cent.

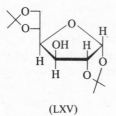

(LXV)

Carbohydrates appear to be very effective as chiral ligands for Mg^{2+}, and by use of the α-D-glucofuranose derivative (LXV) optical yields as high as 70 per cent have been obtained in reactions between ketones and Grignard reagents, e.g. between methylmagnesium bromide and phenylcyclohexyl ketone.

Stereoselectivity is also observed in reactions of carbonyl compounds that contain a co-ordinating substituent at an adjacent chiral centre, and that can react with Grignard reagents or organolithium compounds by first forming a five- or six-membered ring chelate in which both the substituent and the carbonyl oxygen are co-ordinated with the metal. As a result of acting as a

chelating agent the organic ligand is held fairly rigidly in one specific conformation, and the co-ordinated carbonyl group is therefore attacked selectively from one particular direction. When (±)-3-hydroxy-3-phenylbutan-2-one
is treated with an excess of phenyl-lithium, the diol produced contains a
mixture of the *meso*- and (±)-diastereomers in the ratio of 1 : 7. This is consistent with the initial five-membered ring chelate being preferentially attacked
by phenyl-lithium from the least hindered side of the ring, e.g. the enantiomer
(LXVI) ultimately gives (LXVII) (see also p. 79).

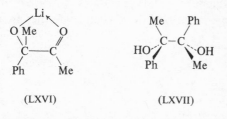

(LXVI)   (LXVII)

When the carbonyl group and the adjacent co-ordinating substituent are in
fairly rigid molecules, the degree of stereoselectivity can be very high. The
reaction between phenylmagnesium bromide and the ketone (LXVIII), for
example, gives almost exclusively the enantiomer (LXIX).

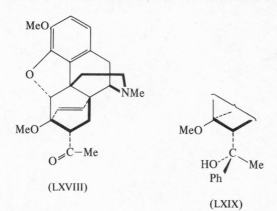

(LXVIII)

(LXIX)

The same principle is involved in the completely stereospecific formation
of the cyclopentane diols (LXXI) by the action of an excess of Grignard
reagents on β-hydroxycyclopentanones of type (LXX). The initial chelate is

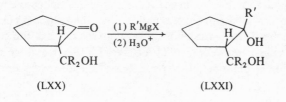

(LXX)   (LXXI)

attacked by the Grignard reagent from the side of the cyclopentanone ring opposite to the CR_2 . OMgBr group, so that the chelate ring may remain intact with both oxygens still co-ordinated with the same magnesium atom.

4.9. Intramolecular additions to carbonyl compounds

On many occasions it has been suggested that in the reduction of aldehydes and ketones by reagents such as metal alkoxides, Grignard reagents, aluminium alkyls and aminolithium compounds, a hydride anion is transferred from the reductant to the carbonyl compound *via* a metal complex formed from the two reactants.

For example, in the Meerwein–Pondorf reduction of carbonyl compounds by aluminium isopropoxide, or the more recently introduced reagent lithium isopropoxide, the co-ordination of the ketone with the metal is thought to be followed successively by intramolecular hydride transfer, separation of the ketone from the resultant complex, and then alcoholysis of the mixed alkoxide.

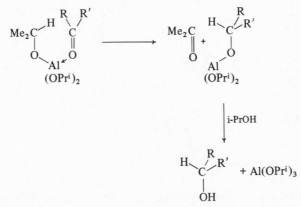

In agreement with this mechanism the use of isopropoxide labelled at the α-position with deuterium gives alcohols deuteriated only at the α-position.

Kinetic studies have shown that in the reduction of acetophenone to 1-phenylethanol the alcohol is formed less rapidly than the ketone is consumed. This suggests that the alcoholysis step is slower than the hydride transfer. The reader is referred to articles by Bradley (1960), Wilds (1944), and Morrison and Mosher (1971), for more detailed discussions of the Meerwein–Pondorf reduction.

An analogous mechanism for the reduction of carbonyl compounds by Grignard reagents has received support from stereochemical studies which were made by Mosher and which have been discussed by Boyd and McKervey (1968) and Morrison and Mosher (1971). In these studies it was observed, for example, that in the reduction of acetophenone by the Grignard reagent obtained from (S)-2-methyl-1-chlorobutane, the 1-phenylethanol formed contained a preponderance of (S)-enantiomer. This is in agreement with the reaction proceeding through a complex (LXXII) whose stereochemistry

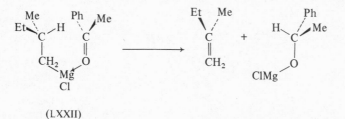

(LXXII)

is such that steric interactions between the two larger groups, phenyl and ethyl, are minimised. Also consistent with initial complex formation is the fact that the reduction of ketones by Grignard reagents is inhibited by Lewis acids such as magnesium iodide that can effectively compete with the Grignard reagent for the ketone.

Complex formation is also the initial step in a number of closely related mechanisms which have been suggested to account for some of the results observed in reactions between carbonyl compounds and reagents such as enolate anions and allylic Grignard reagents. These mechanisms may be summarised by the general equation 4.7.

[4.7]

In the case where X—M is CH_2MgBr, this cyclic mechanism explains why the reaction of a 2-butenyl Grignard reagent with a sterically unhindered ketone affords the alcohol that would be expected from the isomeric Grignard reagent (LXXIII), even though n.m.r. studies show that in ether the two Grignard reagents are in equilibrium and the primary species (LXXIV) constitutes about 99 per cent of the equilibrium mixture (see p. 216).

An identical mechanism also explains the analogous formation of 'abnormal' alcohols from carbonyl compounds and benzylic Grignard reagents. For example, the 76 per cent yield of the diol (LXXVIII) obtained by the action of benzylmagnesium bromide on acetaldehyde (2 equivalents) may be rationalised in terms of the complexes (LXXVI) and (LXXVII).

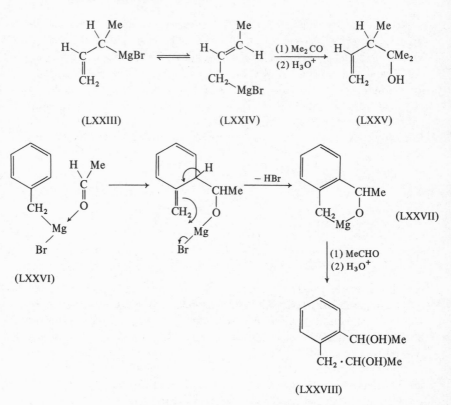

(LXXIII)   (LXXIV)   (LXXV)

(LXXVI)   (LXXVII)

(LXXVIII)

In this connection it is significant that benzylic Grignard reagents, which when co-ordinated with carbonyl compounds give complexes in which the carbonyl carbon atom cannot be held close to the *ortho*-position, as in (LXXIX), give only 'normal' products. Thus, the Grignard reagent derived from *o*-methoxybenzyl bromide reacts with acetaldehyde to give exclusively the alcohol (LXXX).

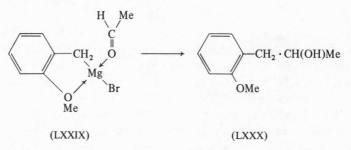

(LXXIX)   (LXXX)

However, doubts have been expressed as to whether the mechanism [4.7] really operates in the reaction between carbonyl compounds and allylic Grignard reagents, on the grounds that in systems where diastereomeric

products are formed, the degree of stereoselectivity is not as high as one
would predict from the cyclic mechanism. The reaction of 2-butenylmagnesium
bromide and acetaldehyde gives an equal mixture of the *threo*- and *erythro*-
forms of the alcohol (LXXXII), whereas if the cyclic mechanism were operating
a high proportion of the *threo*-form would be expected because of the tendency
of the reaction to proceed *via* the complex (LXXXI) in which interactions
between the two methyl groups are minimised.

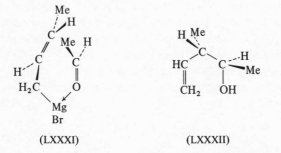

(LXXXI)   (LXXXII)

It has therefore been suggested that the actual mechanism is a bimolecular
electrophilic process (S_E2'), whose stereochemical requirements are not as
critical as those of the cyclic mechanism (see (LXXXIII)).

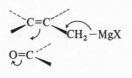

(LXXXIII)

In contrast, aldol condensations between lithium enolates and aldehydes or
ketones often exhibit a high degree of stereoselectivity, and it is claimed that
this is due to the reaction proceeding by the cyclic mechanism [4.7; X = O].
Thus, the 95 per cent yield of the diastereomer (LXXXV) obtained from the
reaction between the lithium enolate of cyclopentanone and isobutyraldehyde
in weakly solvating media is thought to be the direct result of the reaction
proceeding largely *via* the complex (LXXXIV) in which steric repulsions are
minimal.

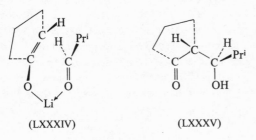

(LXXXIV)   (LXXXV)

The Reformatskii reaction provides another example for which the cyclic mechanism has been advocated. Until recently it was generally thought that the reactive intermediate generated in this reaction was similar to a Grignard reagent and contained a C—Zn bond, but infrared and n.m.r. studies have shown that a structure of type (LXXXVI) only exists in strongly co-ordinating solvents such as DMSO and HMPTA. It is now known that in the solvents traditionally used in Reformatskii reactions, e.g. ether–benzene mixtures, the active intermediate has the enolic structure (LXXXVII), as shown by the absence of the normal ester carbonyl stretching frequency in the infrared. The n.m.r. spectrum of a benzene solution of the α-lithium derivative of t-butyl acetate is also indicative of the O–metal species $H_2C=C(OLi)(O.Bu^t)$. This derivative, which is a stable solid in the absence of moisture, is a more convenient reagent for carrying out Reformatskii transformations than the organozinc reagent prepared *in situ* from methyl or ethyl bromoacetate.

$$\underset{\underset{R^2}{|}}{\overset{\overset{R^1}{|}}{BrZn-C-CO_2R}}$$

(LXXXVI)

$$\underset{R^2}{\overset{R^1}{\diagdown}}C=C\underset{OZnBr}{\overset{OR}{\diagup}}$$

(LXXXVII)

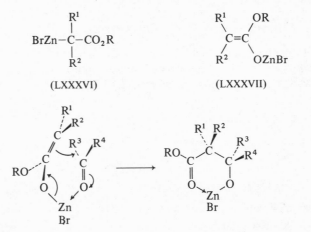

[4.8]

To accommodate these observations, and also the fact that unlike zinc dialkyls the organozinc compounds formed in the Reformatskii reaction react very rapidly with carbonyl compounds, the cyclic mechanism [4.8] has been proposed. In contrast to the analogous mechanism for the reaction of carbonyl compounds and allylic Grignard reagents (see above) this mechanism is supported to some extent by stereochemical results (see Gaudemar (1972) and Rathke (1975) for discussions on this and other aspects of the reaction). For example, in systems that can give rise to diastereomers, e.g. acetophenone and methyl 2-bromobutyrate, the *threo*-form of the hydroxy ester, i.e. the diastereomer predicted from stereochemical considerations of the cyclic mechanism, is the major product (usually 70–80 per cent). It should be noted, however, that the *erythro*-form becomes the major product at low temperatures (~ -70 °C), and when DMSO is used as the solvent.

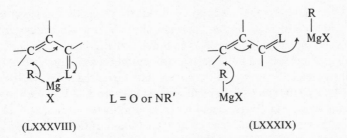

(LXXXVIII)   L = O or NR′   (LXXXIX)

Of the two mechanisms (shown in (LXXXVIII) and (LXXXIX)) proposed for the 1,4-addition of Grignard reagents to αβ-unsaturated carbonyl compounds and imines, the first is analogous to the general type [4.7]. Here again, there is a small amount of stereochemical evidence for the participation of this type of mechanism as, for example, in the reaction involving the imines derived from αβ-unsaturated aldehydes and the t-butyl ester of (L)-t-leucine. This reaction affords (after subsequent hydrolysis) optically active aldehydes having an optical purity of: > 90 per cent, whose configurations are consistent with the participation of magnesium chelates of type (XC) through which the alkyl group of the Grignard reagent adds to the β-carbon from the less hindered side.

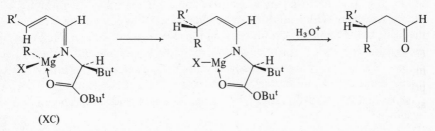

(XC)

The use of copper alkyls or mixtures of Grignard reagents and copper(I) compounds for achieving 1,4-additions appears not to involve the cyclic mechanism shown in (LXXXVIII), see p. 238.

4.10. Formation of cyclopropanes from alkenes

A number of halomethylmetal compounds react stereospecifically with alkenes to form cyclopropanes.

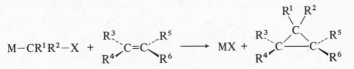

Such organometallics are called 'carbenoids' because in one or two cases there is very strong evidence to show that carbenes are generated from the organometallic during the reaction. The range of metals that form carbenoids is very wide and includes lithium, magnesium, cadmium, aluminium, iron, lead, tin and mercury. Halomethyl and polyhalomethyl compounds of the last element have proved to be particularly useful for preparing cyclopropanes from alkenes, and by means of the appropriate mercurial, CCl_2, $CClBr$, CBr_2, $CClF$, $CHCl$, $CHBr$ and CH_2 have all been added across carbon–carbon double bonds, notably by D. Seyferth and his collaborators (see review, 1972). Of the various mercurials which can cause the addition of CCl_2, phenyltrichloro-methylmercury is exceptionally effective, and has been used for preparing *gem*-dichlorocyclopropanes from simple alkenes, $\alpha\beta$-unsaturated ketones, nitriles, and esters. Even the highly hindered double bond in Δ^7-unsaturated steroids reacts in the expected manner.

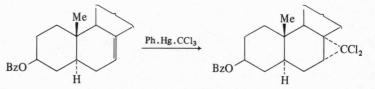

For the addition of CH_2 the Simmons–Smith reaction is preferred, although with alkenes that are prone to cationic polymerisation, e.g. vinyl ethers, treatment with a mixture of diethylzinc and di-iodomethane is recommended. In the Simmons–Smith reaction the alkene reacts with an α-iodo-methylzinc compound generated *in situ* from di-iodomethane and a zinc-copper couple in an ether-containing solvent. For convenience this intermediate zinc compound is usually formulated as $I . Zn . CH_2I$, but it is possible that it is actually $Zn(CH_2I)_2$ complexed with zinc iodide and ether as shown in (XCI).

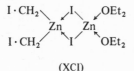

(XCI)

A characteristic of the Simmons–Smith reaction is the directing effect which is exhibited by several types of functional groups when they are close to the double bond in the alkene. For example, with allylic alcohols in which the double bond is part of a cyclopentene or cyclohexene ring, the addition of CH_2 takes place on the side of the ring where the hydroxyl group is with a very high degree of stereoselectivity.

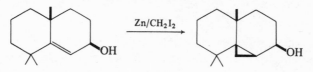

Ether and ester groups have the same effect, and it appears that all these 'directing' substituents operate by co-ordinating with the zinc in the Simmons–Smith reagent, thus ensuring that the reagent attacks that face of the double bond that is nearest to the co-ordinated substituent. This explanation also holds for the reaction with allylic cyclo-octenols and cyclononenols, for these alcohols exist largely in conformations (e.g. (XCII)) that would be expected to cause *trans*-addition of CH_2 with respect to the hydroxyl group, and this is in fact observed.

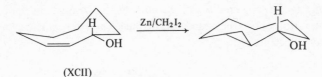

(XCII)

The suggestion by Simmons and Smith that CH_2 is transferred from an iodomethylzinc compound to the double bond of the alkene in a bimolecular one-step process with the transition state (XCIII) has been widely accepted.

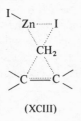

(XCIII)

However, experimental evidence indicates that with some of the mercurials a carbene is initially formed from the organometallic and this then reacts with the alkene to give the cyclopropane. It seems likely that this is also true with a high proportion of the other carbenoid organometallics. This carbene formation can be imagined to take place either by a concerted mechanism with a three-membered ring transition state (XCIV) (as seems probable with some of the mercurials) or by stepwise processes *via* carbene complexes.

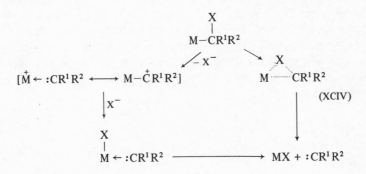

With some carbenoids the reaction with the alkene may involve these intermediate carbene complexes rather than the free carbene produced from them.

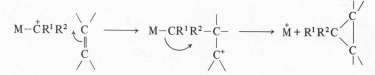

Similar metal–carbene complexes are probably also involved in the copper-catalysed formation of cyclopropanes from alkenes and diazoalkanes or α-diazocarboxylic esters.

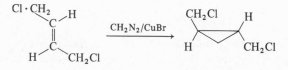

A substantial degree of asymmetric synthesis can be induced into the last type of addition (which has been briefly reviewed by Cowell and Ledwith, 1970) by using metal complexes of optically active ligands to promote the decomposition of the diazo compound. Typical of the complexes which have been used are [(−)-tribornyl phosphite] copper(I) chloride (XCV) and bis-[(+)-camphorquinonedioximato] cobalt(II) (XCVI). The same effect can be achieved

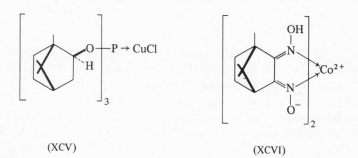

(XCV)   (XCVI)

in the Simmons–Smith reaction by incorporating into the reaction mixture optically active alcohols, such as (−)-menthol, which can co-ordinate with the zinc in the intermediate iodomethylzinc compound. The use of (−)-menthol as the alcohol component of the ester group in an αβ-unsaturated ester also causes a small degree of assymmetric synthesis in the Simmons–Smith addition to the double bond.

5 Oxidative addition and insertion

5.1. Oxidative addition

In §1.6 it was pointed out that in many complexes formed from metals in
low oxidation states and alkenes or related unsaturated ligands there is a very
high degree of back-bonding from the metal, and such complexes can justifiably be represented by valence-bond structures in which the metal has
formally undergone a two-electron oxidation. The complexes formed from
bis(triphenylphosphine)platinum(0) and tetracyanoethylene and carbon
disulphide, for example, can be represented by the structures (I) and (II)
respectively.

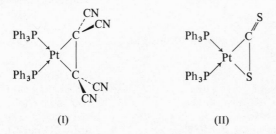

(I)   (II)

Accordingly, the addition of the ligand to the metal is termed an oxidative
addition, while the reverse process, i.e. dissociation of the ligand from the
metal, is termed a reductive elimination. In such oxidative additions the
bond order of the functional group in the unsaturated ligand which is added is
formally reduced by one unit. A logical extension of this is the cleavage of a
ligand into two parts by the complete fission of a single bond.

$$M: + L{-}L' \rightarrow M\overset{\displaystyle L}{\underset{\displaystyle L'}{<}}$$

Many single bonds can in fact be cleaved in this manner by metal complexes in which the metal is in a low oxidation state, and it is this type of

oxidative addition that is discussed in this section. Examples of different types of bonds that can be cleaved by means of an oxidative addition are given in table 5.1.

Many organic reactions of industrial importance that use transition-metal compounds involve an oxidative addition as a key step. Because of this,

Table 5.1. *Oxidative additions of the type:*

$$M + L-L' \rightarrow L-M-L'$$

Bond type	Compound	M	Products
L—L'			
C—C	Ph_3C-CPh_3	$Ni(COD)_2$	$Ni(CPh_3)_2$ + 2COD
	$NC-CN$	$Pd(PPh_3)_4$	$Pd(CN)_2(PPh_3)_2$ + 2PPh_3
	$Ph-CN$	$Ni[Et_2P(CH_2)_2PEt_2]_4$	$NiPh(CN)[Et_2P(CH_2)_2PEt_2]_2$ + 2 $Et_2P(CH_2)_2PEt_2$
	$n\text{-}C_6H_{13}-\triangleleft$	$PtCl_2(C_2H_4)_2$	$n\text{-}C_6H_{13}-\triangleleft PtCl_2$ + $2C_2H_4$
C—X	$n\text{-}Bu-I$	$[Rh(Bu^tNC)_4]^+$	$[RhBu^nI(Bu^tNC)_4]^+$
	$H_2C=CH-Cl$	$Pd(PMePh_2)_4$	$PdCl(H_2C=CH)(PMePh_2)_2$ + 2PMePh_2
	$Ph-Cl$	$Ni(PPh_3)_4$	$NiClPh(PPh_3)_2$ + 2PPh_3
	[benzothiazol-2-yl]—Cl	$IrCl(CO)(PMe_2Ph)_2$	[benzothiazol-2-yl]—$IrCl_2(CO)(PMe_2Ph)_2$
	$PhCO-Cl$	$Pd(PPh_3)_4$	$PdCl(COPh)(PPh_3)_2$ + 2PPh_3
	$EtO_2C.CO-Cl$	$Pt(PEt_3)_4$	$PtCl(CO.CO_2Et)(PEt_3)_2$ + 2PEt_3
C—O	$Me.CO-O.CO.Me$	$Pt(PPh_3)_4$	$Pt(CO.Me)(O.CO.Me)(PPh_3)_2$ + 2PPh_3
C—S	$Me_2N.CS-SMe$	$RhCl(PPh_3)_3$	$RhCl(SMe)(CS.NMe_2)(PPh_3)_2$ +PPh_3
C—H	$PhC\equiv C-H$	$Pd(PPh_3)_4$	$PdH(C\equiv CPh)(PPh_3)_2$ + 2PPh_3
	$N\equiv C-H$	$Pt(PPh_3)_4$	$PtH(CN)(PPh_3)_2$ + 2PPh_3
O—H	$CF_3.CO.O-H$	$IrCl(PPh_3)_3$	$IrClH(O.CO.CF_3)(PPh_3)_2$ + PPh_3
S—H	$o\text{-}MeC_6H_4.S-H$	$IrCl(PPh_3)_3$	$o\text{-}Me.C_6H_4.S\text{-}IrClH(PPh_3)_2$ + PPh_3
N—H	[phthalimide] N—H	$Pt(PPh_3)_4$	[phthalimide] N—$PtH(PPh_3)_2$ + 2PPh_3
X—H	$Cl-H$	$IrCl(CO)(PPh_3)_2$	$IrCl_2H(CO)(PPh_3)_2$
H—H	$H-H$	$[Rh(PBu^n_3)_4]^+$	$[RhH_2(PBu^n_3)_4]^+$
O—O	Bu^tO-OBu^t	$Ni(COD)_2$	$Ni(OBu^t)_2$ + 2COD
S—S	$MeS-SMe$	$Ti(\eta\text{-}C_5H_5)_2(CO)_2$	$Ti(\eta\text{-}C_5H_5)_2(SMe)_2$ + 2CO
X—X	$Br-Br$	$Fe(CO)_5$	$FeBr_2(CO)_4$ + CO

and also because of the close similarity between the oxidative addition of simple covalent molecules such as dihydrogen and dioxygen and their chemisorption (and associated activation) on transition-metal surfaces, a large number of studies on oxidative addition have been made (see reviews by Collman and Roper (1968), Halpern (1970), and Deeming (1972)). Most of the studies have concerned oxidative additions to complexes of metals with either a d^8 or a d^{10} electronic configuration, i.e. $d^8 \rightarrow d^6$ and $d^{10} \rightarrow d^8$ transformations. Of the metals of interest to the organic chemist, iron(0), rhodium(I) and platinum(II) are examples of the d^8 configuration, while palladium(0), nickel(0), copper(I) and silver(I) are examples of d^{10}. With metals having a d^8 configuration the tendency to undergo oxidative addition increases as one passes from first- to third-row transition elements, and as the formal oxidation state of the metal decreases. In Group VIII the tendency also increases as one passes from right to left, and the approximate order of reactivity $Os(0) > Ru(0) \sim Ir(I) > Fe(0) \sim Rh(I) > Pt(II) > Co(I) \sim Pd(II) > Ni(II)$ has been assigned. The reactivity of any particular metal, however, is highly dependent upon the ligands co-ordinated with that metal, and a comparison between the reactivities of two metals with different ligands is of very limited value. In general, ligands that decrease the electron density on the metal decrease the reactivity of the metal with respect to oxidative addition. Metal complexes of aryl phosphines are therefore less reactive towards oxidative addition than the corresponding complexes of alkyl phosphines. Carbon monoxide is deactivating compared with triaryl phosphines because of the stronger π-accepting properties of this ligand, and the halogens exhibit an order of influence which correlates with their σ-donor strengths. The equilibrium constants for the formation of the dihydride species in equation 5.1 for example, decreases in the order $I > Br > Cl$.

$$\begin{array}{ccc} \underset{X}{\overset{Ph_3P}{\diagdown}}\!\!\!\underset{\diagup}{\overset{\diagup}{Ir}}\!\!\!\underset{PPh_3}{\overset{CO}{\diagup}} \; + H_2 & \rightleftharpoons & \underset{X}{\overset{Ph_3P}{\diagdown}}\!\!\!\underset{\overset{|}{CO}}{\overset{H}{Ir}}\!\!\!\underset{PPh_3}{\overset{H}{\diagup}} \end{array} \qquad [5.1]$$

An additional complication in comparing the reactivities of metal complexes towards oxidative addition is that with some complexes the dissociative loss of a ligand to produce a co-ordinatively unsaturated species (see p. 2) is a prerequisite for the addition to occur. Thus the highly reactive species, bis(triphenylphosphine)platinum(0), which undergoes oxidative additions with a very wide range of saturated and unsaturated compounds, is produced by a dissociative mechanism from the tetrakis complex.

$$Pt(PPh_3)_4 \; \underset{}{\overset{-\,PPh_3}{\rightleftharpoons}} \; Pt(PPh_3)_3 \; \underset{}{\overset{-\,PPh_3}{\rightleftharpoons}} \; Pt(PPh_3)_2 \; \overset{L-L'}{\longrightarrow} \; \underset{Ph_3P}{\overset{Ph_3P}{\diagdown}}\!\!\!\underset{\diagup}{\overset{\diagdown}{Pt}}\!\!\!\underset{L'}{\overset{L}{\diagup}}$$

In this case the initial dissociation is complete in benzene at room temperature, but the second dissociation has an equilibrium constant of about 10^{-4}.

One route by which five-co-ordinate d^8 complexes undergo oxidative addition is by initial dissociation of one of the ligands to give a four-co-ordinate species which then oxidatively adds the addendum.

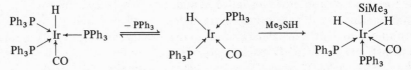

In oxidative additions involving the cleavage of a single bond in the addendum, both *cis-* and *trans-*additions to the metal have been observed. Thus stereospecific *trans-*addition of methyl bromide or methyl iodide to the iridium(I) complex (III; L = PMePh$_2$) occurs, but hydrogen bromide or hydrogen iodide in dry benzene gives *cis-*addition. In both cases the stereochemistry of the product is kinetically controlled.

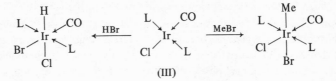

(III)

Undoubtedly the stereochemistry of any oxidative addition is governed in the first place by the actual mechanism of the addition, but it has been shown that in some cases the initial product may subsequently isomerise to give a more stable stereoisomer. Except in one or two specific cases, however, little work has been done on the mechanism of oxidative addition. In some cases, such as the addition of methyl iodide to IrCl(CO)(PPh$_3$)$_2$, an S_N2 type of mechanism similar to the nucleophilic displacements observed with anionic iron(0) and cobalt(I) complexes (see p. 41) operates.

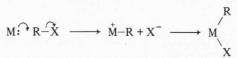

With some five-co-ordinate d^8 complexes an intermediate, e.g. (IV, M = Ir or Rh), consistent with this mechanism has been isolated.

$$[M(PPh_3)_2(CN \cdot Bu^t)_3]^+ \xrightarrow{\text{MeI}} [M(Me)(PPh_3)_2(CN \cdot Bu^t)_3]^{2+} + I^-$$

(IV)

$$[M(I)(Me)(PPh_3)(CN \cdot Bu^t)_3]^{2+}$$
$$+ PPh_3$$

However, additions of primary alkyl and cycloalkyl bromides to certain four-co-ordinate d^8 complexes and to d^{10} complexes of platinum(0) with triaryl or trialkylphosphine ligands proceed by a radical mechanism of the following type.

$$M: + R-X \rightarrow \dot{M}-X + R^{\cdot} \rightarrow M\diagup^{R}_{\diagdown X}$$

Thus the addition of n-butyl bromide to *trans*-IrCl(CO)(PMe$_3$)$_2$ is accelerated by small quantities of oxygen and radical initiators, but is inhibited by radical scavengers. With halides in which the halogen is bonded to an asymmetric carbon the addition is accompanied by racemisation. In some of these radical additions the capture of the radical (R·) by the species ·M−X occurs sufficiently slowly for these two radicals to undergo alternative reactions, and products other than that of the expected oxidative addition are then observed. The reaction in toluene of cyclohexyl bromide with Pt(PEt$_3$)$_2$ formed by dissociation from the tris(triethylphosphine) complex, for example, affords a good yield of *trans*-PtBr(benzyl)(PEt$_3$)$_2$ because of the participation of benzyl radicals generated by the interaction of cyclohexyl radicals with the solvent.

Several groups of workers have examined the mechanism of the formation of σ-bonded alkenylmetal complexes by the oxidative addition of alkenyl halides to certain zerovalent metal complexes, notably phosphine complexes of Pt(0), Pd(0) and Ni(0). Examples of such additions are the formation of *trans*-PtCl(CF=CF$_2$)(PPh$_3$)$_2$ and *trans*-PdCl(CH=CHCl)(PPh$_2$Me)$_2$ from the reaction of the appropriate alkenyl chloride with Pt(PPh$_3$)$_4$ and Pd(PPh$_2$Me)$_4$ respectively. There is convincing evidence that these reactions involve the initial formation of an alkene π-complex – presumably after prior dissociation of two of the phosphine ligands – which then isomerises to the σ-bonded complex.

$$Pt(PPh_3)_4 \xrightleftharpoons{-2PPh_3} Pt(PPh_3)_2 \xrightarrow{F_2C=CF_2} \left[\begin{array}{c} \text{Ph}_3\text{P} \diagdown \quad \diagup \text{CF}_2 \\ \text{Pt} \leftarrow \| \\ \text{Ph}_3\text{P} \diagup \quad \diagdown \text{CF}_2 \end{array} \longleftrightarrow \begin{array}{c} \text{Ph}_3\text{P} \diagdown \quad \diagup \text{CF}_2 \\ \text{Pt} \diagup | \\ \text{Ph}_3\text{P} \diagup \quad \diagdown \text{CF}_2 \end{array} \right]$$

$$\downarrow$$

$$\begin{array}{c} \text{F} \diagdown \quad \diagup \text{PPh}_3 \\ \text{Pt} \\ \text{Ph}_3\text{P} \diagup \quad \diagdown \text{CF=CF}_2 \end{array}$$

Indeed, in the formation of σ-complexes such as PtCl(CCl=CCl$_2$)(PPh$_3$)$_2$ the π-complex has been isolated and found to isomerise as proposed when heated. The observation that the stereochemistry of the original alkene is retained in the final σ-bonded complex, e.g. ClHC$\stackrel{cis}{=}$CHCl gives Cl—M—CH$\stackrel{cis}{=}$CHCl, suggests that the initial π-complex rearranges by a concerted, conrotatory process (see (V)) or *via* a tight ion-pair (VI) containing halide anion and a metal-stabilised carbonium ion. The latter possibility is consistent with the loss of stereospecifity that occurs when the oxidative addition is carried out in the presence of Ag⁺, for this species would be expected to destroy the intimate nature of the ion-pair by capturing the halide anion.

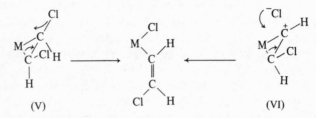

(V)   (VI)

Besides their preparation by the oxidative addition of alkenyl halides, organometallics of the type X.M.CR=CHR′ may also be prepared by the action of protonic acids (HX) on π-complexes of alkynes with metals in low oxidation states. For example, the complex formed from diphenylacetylene and bis(triphenylphosphine)platinum(0) reacts with hydrogen chloride to give the σ-alkenylplatinum(II) compound (VII). This type of reaction probably involves an oxidative addition of H—X to the acetylene complex followed by a reductive elimination. This pathway accounts for the fact that in the product the metal is on the same side of the double bond as the proton supplied by the acid.

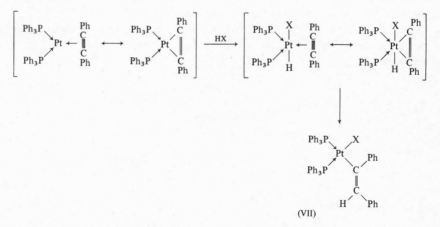

(VII)

The formation of σ-bonded alkenyl organometallics from π-complexes of alkynes is of interest in connection with the mechanism of the hydrocarboxylation of alkynes (see p. 288).

There are a number of oxidative additions in which the carbon–halogen bond of an aryl halide is cleaved.

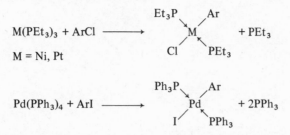

It has been proposed that when the aryl halide bears electron-withdrawing substituents, these additions proceed by a conventional type of nucleophilic displacement at unsaturated carbon with the metal functioning as the nucleophile.

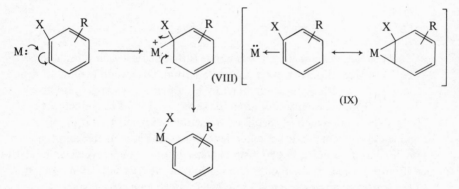

However, the close relationship between the ionic intermediate (VIII) of such a process and the bicyclic complex (IX), which is directly analogous to the π-complex involved in the oxidative addition of an alkenyl halide, suggests that the bicyclic complex may be an intermediate in the reaction.

The mechanisms by which organic halides oxidatively add to transition metals in Group VIII have been discussed by Stille and Lau (1977).

In some oxidative additions a carbon–hydrogen bond is cleaved by a metal in a low oxidation state. In a few cases the addition is intermolecular, as in the reaction between benzene and bis(π-cyclopentadienyl)tungsten (X), but most of the additions are intramolecular and the carbon–hydrogen bond is in one of the ligands attached to the metal, usually through nitrogen or phosphorus. A typical example is the formation of the iridium(III) species (XII) when the complex (XI) is heated in cyclohexane (see p. 230).

Examples of ligands that contain C—H bonds capable of undergoing an intramolecular oxidative addition with a transition-metal ion (Pd, Pt, Ir and Rh are the metals that have been studied most frequently) are shown in structures (XIII)–(XVI), together with the particular C—H bond that is involved in the reaction.

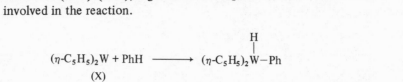

$$(\eta\text{-}C_5H_5)_2W + PhH \longrightarrow (\eta\text{-}C_5H_5)_2\overset{\displaystyle H}{\underset{}{W}}-Ph$$

(X)

IrCl(PPh₃)₃ ⟶ (XII)

(XI)

(XIII)   (XIV)   (XV)

$H \cdot CH_2 \cdot CMe_2 \cdot CH_2 \cdot PBu_2^t$

(XVI)

One characteristic of these C—H bonds is that deuteriation is usually effected very readily by contact with dideuterium (D₂) when the parent ligand is co-ordinated with a transition-metal hydride that can undergo oxidative additions. Thus all twenty-four aromatic *ortho*-hydrogens in the complex $H_2Ru(PPh_3)_4$ are selectively replaced by deuterium when a solution of this complex in benzene is stirred under deuterium gas. These deuteriations may be visualised as proceeding in two distinct steps both of which involve an oxidative addition followed by a reductive elimination. In the first step an M—H bond in the starting complex is replaced by M—D, and in the second step a C—H bond in the ligand is replaced by C—D.

$$M-H + D_2 \rightleftharpoons \overset{\displaystyle D}{\underset{}{D-M-H}} \rightleftharpoons M-D + HD$$

$$M-D + H-\overset{}{\underset{}{C}}{-} \rightleftharpoons H-\overset{\displaystyle D}{\underset{}{M}}-\overset{}{\underset{}{C}}{-} \rightleftharpoons M-H + D-\overset{}{\underset{}{C}}{-}$$

Mechanistically these deuteriations are probably closely related to similar selective deuteriations which have been observed under heterogeneous conditions, e.g. the exchange of the *ortho*-hydrogens of aniline and phenol by the action of D_2O in the presence of nickel-kieselguhr.

The oxidative-addition of a C—H bond results in a formal two-electron oxidation of the metal concerned. In a number of systems, however, the metal hydride which is formed undergoes a reductive elimination of the hydride ligand and one of the ligands that was present in the starting complex, and the overall transformation does not therefore lead to an increase in the oxidation state of the metal.

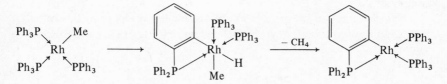

Although a two-step mechanism of this type may be visualised for a high proportion of those reactions in which an organic ligand is metallated at a specific position by a transition metal, e.g. the metallation of azobenzene by sodium chloropalladate(II) to give the halogen-bridged dimer of the complex (XVII, X = N), the actual mechanism of these metallations is still the subject of debate. Indeed, it is possible that some of them do not involve oxidative additions at all. However, as already indicated (p. 175) in connection with the process of lithiation, regiospecific metallations are of synthetic interest because they can provide a route for the indirect introduction of various functional groups into a specific position of an organic ligand. For example, if the complex (XVII; X = N) is treated with $Na^+ [Co(CO)_4]^-$ the Pd—Cl

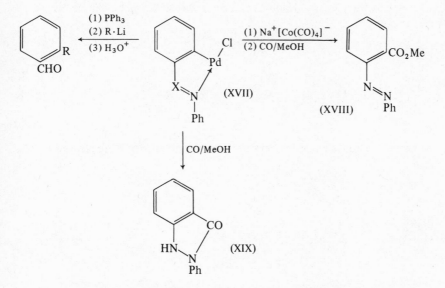

entity is replaced by $Co(CO)_3$, and the resultant cobalt complex can be carbonylated in methanol to afford the methyl ester (XVIII). Direct carbonylation of the palladium complex (XVII; X = N) in methanol gives an excellent yield of 2-phenyl-3-indazolinone (XIX). The analogous complex (XVII;

X = CH) obtained by metallation of the parent Schiff base with Pd(OAc)$_2$/ NaCl, may readily be converted into *ortho*-alkyl substituted benzaldehydes as indicated, thus providing an exceptionally convenient route to these compounds.

For a discussion of the metallation of organic ligands by transition metals, the reader is referred to reviews by Dehand and Pfeffer (1976) and Bruce (1977).

5.2. Oxidative addition in organic synthesis

A number of organic reactions that are catalysed by transition-metal compounds involve oxidative additions and reductive eliminations combined with insertions of organic molecules into metal–hydrogen or metal–carbon bonds. Catalytic hydrogenation (§5.5) and hydroformylation (p. 283) fall into this category. It is evident from current studies in organometallic chemistry, however, that even a simple combination of oxidative addition and reductive elimination is synthetically important in that it provides methods of joining together different organic groups that are capable of being individually bonded to a metal.

In the simplest reaction scheme an addendum $L^1 - L^2$ is oxidatively added to a metal which bears a ligand L, and the resultant complex eliminates the species $L-L^1$.

$$M-L + L^1-L^2 \rightarrow \overset{\overset{\displaystyle L^1}{|}}{L^2-M-L} \rightarrow L^2-M + L-L^1 \qquad [5.2]$$

This is the route by which certain transition-metal alkyls are cleaved by acids.

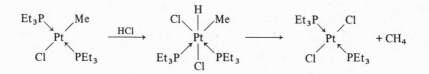

Of greater interest to the organic chemist, however, are the specific examples of the linking of two aromatic residues and the formation of esters *via* intermediate platinum(IV) and cobalt(III) complexes respectively.

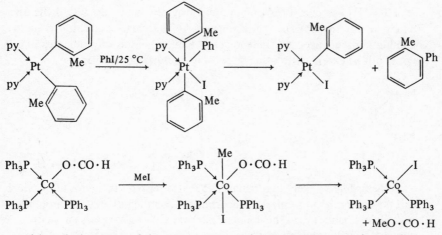

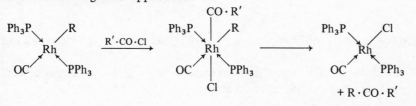

Although the scope of these two reactions has not yet been explored, the use of rhodium(I) in a synthesis of mixed ketones from acid chlorides has been shown to be of general application.

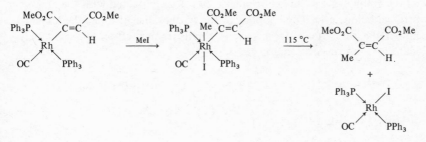

This synthesis, which tolerates the presence of CHO, CO_2R and CN groups, has recently been used for preparing (S)-4-methylheptan-3-one, the principal alarm pheromone of fungus-growing ants of the genus *Atta*, from (S)-2-methylpentanoyl chloride.

$$RhCl(CO)(PPh_3)_2 \xrightarrow{EtLi} RhEt(CO)(PPh_3)_2 \longrightarrow$$

Rhodium(I) has also been used in a general, stereospecific synthesis of alkenes, for example.

1,1,1-^{2}H$_3$ acetone, a compound previously obtained only with difficulty and in an impure form, has been prepared by an oxidative addition/reductive elimination sequence, and there appears to be no reason why this preparation could not be extended to other selectively deuteriated ketones.

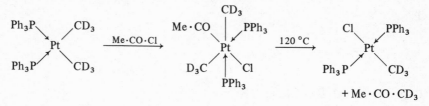

$$+ Me \cdot CO \cdot CD_3$$

It seems highly likely that equation 5.2 also represents the route by which carbon–carbon bond formation occurs in the coupling of alkynyl, aromatic and alkenyl halides with the copper(I) derivatives of alk-1-ynes and with copper(I) perfluoroalkyls.

$$PhI + n\text{-}C_7F_{15} \cdot Cu \quad \rightarrow \quad n\text{-}C_7F_{15}\overset{\overset{\displaystyle Ph}{|}}{-}Cu-I \quad \rightarrow \quad n\text{-}C_7F_{15} \cdot Ph + CuI$$

$$p\text{-}Me \cdot C_6H_4I + Ph \cdot C{\equiv}C \cdot Cu \quad \rightarrow \quad PhC{\equiv}C\overset{\overset{\displaystyle p\text{-}Me \cdot C_6H_4}{|}}{-}Cu-I \quad \rightarrow \quad p\text{-}Me \cdot C_6H_4 \cdot C{\equiv}C \cdot Ph + CuI$$

The same process could also be involved in the Ullman reaction for preparing biaryls from aryl halides and metallic copper.

$$ArI \xrightarrow{Cu} ArCu \xrightarrow{Ar'I} Ar\overset{\overset{\displaystyle Ar'}{|}}{-}Cu-I \longrightarrow Ar-Ar' + CuI$$

Another type of organic reaction for which the oxidative addition/reductive elimination sequence shown in equation 5.2 has been proposed is the metal-promoted coupling of allylic halides to give 1,5-dienes. Although this process is best known as a side-reaction in the preparation of allylic Grignard reagents it can be made preparatively important if allylic halides are treated with suitable inorganic reagents. Iron powder in dimethylformamide was formerly recommended, but has now been replaced by nickel tetracarbonyl which can be used for coupling allylic acetates and tosylate as well as allylic halides. In all the couplings, mixtures of isomeric dienes are obtained with the major product being derived by coupling of the two residues through the least hindered carbon of the allylic system. Allylic isomers afford approximately identical mixtures of isomeric dienes.

$$\left. \begin{array}{c} Me_2C{=}CH \cdot CH_2Cl \\ \text{or} \\ Me_2CCl-CH{=}CH_2 \end{array} \right\} \xrightarrow{Ni(CO)_4} \begin{array}{cc} Me_2{=}CH \cdot CH_2 \cdot CH_2 \cdot CH{=}CMe_2 & 65 \text{ per cent} \\ + & \\ Me_2{=}CH \cdot CH_2 \cdot CMe_2 \cdot CH_2 {=}CH_2 & 35 \text{ per cent} \end{array}$$

Intramolecular coupling to give cyclic 1,5-dienes is best accomplished by the action of nickel tetracarbonyl on bis(allylic bromides) in an aprotic dipolar solvent such as dimethylformamide or *N*-methylpyrrolidone. This coupling technique has been used in a preparation of several macrocyclic dienes including the 4,5-*cis*-isomer (XX) of humulene (XXI) and the [12.1.0]-bicyclic diterpene casbene (XXII).

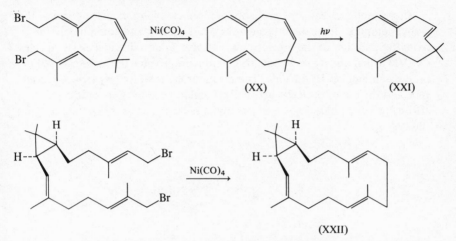

(XX)   (XXI)

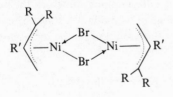

(XXII)

It appears that in the coupling of allylic halides with nickel tetracarbonyl, the initial stage of the reaction is the oxidative addition of the halide to the zerovalent nickel to give a π-allylnickel(II) halide which can be isolated as a stable dimer if benzene is used as the solvent; e.g. allyl bromide affords (XXIII; R = R' = H).

(XXIII)

In dimethylformamide these π-allyl complexes react with alkyl, allyl, alkenyl and aryl iodides or bromides (generally alkyl halides are the least reactive) to give good yields of cross-coupled products in which the allylic residue (if unsymmetrical) is linked to the other organic residue through its least hindered carbon, e.g. the dimeric complex (XXIII, R = Me, R' = H) prepared from 3,3-dimethylallyl bromide reacts with the iodide (XXIV) to give α-santalene (XXV). Use of the π-allyl complex (XXIII; R = H, R' = OMe) prepared from 1-bromo-2-methoxyprop-2-ene in the coupling reaction provides a means of replacing a halogen by either a $CH_2 \cdot C(OMe) = CH_2$ or (after mild acid hydrolysis) a $CH_2 \cdot CO \cdot Me$ group.

(XXIV)   (XXV)

These coupling reactions probably proceed *via* the monomeric form of the
π-allyl complex rather than through the actual dimer, and could involve an
oxidative addition of the organic halide to give an unstable nickel(IV) species
(XXVI) which undergoes a reductive elimination to give the cross-coupled
product and nickel(II) dihalide. The reactions are inhibited by radical-scaven-
gers, and the use of optically active alkyl halides, e.g. (+)-2-iodo-octane,
affords racemic products, suggesting that a radical process is involved at some
stage.

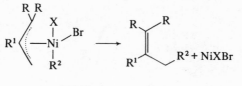

(XXVI)

Useful reviews on the coupling of allylic halides by means of π-allyl complexes are those by Semmelhack (1972), Baker (1973) and Noyori (1976).

So far the reactions described can be rationalised by the scheme shown in
equation 5.2, i.e. an oxidative addition followed by a reductive elimination.
In some reactions, however, the reductive elimination appears to occur
within a species formed by a nucleophilic displacement in which the nucleophile is a metal complex with the metal in a low oxidation state. When this
metal complex is a singly charged anion this variation can be represented by
equation 5.3.

$$L-M \quad L'-X \xrightarrow{\;-X^-\;} L-M-L' \longrightarrow M: + L-L' \qquad [5.3]$$

Although many nucleophilic substitutions by anionic metal complexes
are known (see p. 41) there are comparatively few model systems for the
combined substitution/reductive elimination represented by equation 5.3.
One example involves an intermediate gold(III) complex.

$$Li^+Me_2(PPh_3)Au^- + Me-Br \xrightarrow{\;-LiBr\;} \underset{Ph_3P}{\overset{Me}{\diagdown}}Au\underset{Me}{\overset{Me}{\diagup}} \longrightarrow Ph_3P \rightarrow Au-Me + Me \cdot Me$$

It is possible, however, that equation 5.3 represents the sequence by
which coupled products are formed when alkyl halides react with lithium

di(organo)cuprates. These copper(I) compounds are prepared *in situ* by the addition of copper(I) iodide to two equivalents of the appropriate organo-lithium reagent, and for convenience they are formulated as monomers, $(R_2Cu)^-Li^+$, even though in the solid state – and possibly also in solution – they are oligomeric in nature. Organic halides react with these organocuprates to give cross-coupled products, usually in moderate to excellent yield, and probably *via* an intermediate copper(III) complex.

$$(R_2Cu)^-Li^+ + R'X \rightarrow R-\overset{\overset{\displaystyle R}{|}}{Cu}-R' \rightarrow R-Cu + R-R'$$

The scope of the reaction is very wide, for the expected cross-coupled products are also obtained when organocuprates prepared from alkenyl- and aryl-lithium compounds react with alkenyl, aryl and acyl halides. As organo-cuprates react quite slowly with ketonic and ester carbonyl groups, these groups do not require protection when they are present in the organic halide.

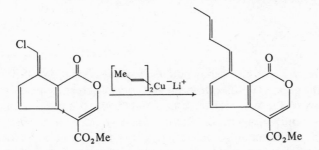

Lithium di(organo)cuprates also react with oxiranes to give the expected alcohol.

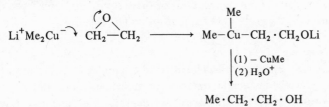

Even with simple oxiranes such as propylene oxide the yields are superior to those obtained by the use of the corresponding Grignard reagent, and it seems highly likely that the reaction proceeds *via* a copper(III) complex formed by a nucleophilic substitution.

$$Li^+Me_2Cu^- \curvearrowright \overset{\diagup O \diagdown}{CH_2-CH_2} \longrightarrow Me-\overset{\overset{\displaystyle Me}{|}}{Cu}-CH_2 \cdot CH_2OLi$$

$$\begin{array}{c} \Big| (1) - CuMe \\ \Big| (2) \, H_3O^+ \end{array}$$

$$Me \cdot CH_2 \cdot CH_2 \cdot OH$$

The idea that the nucleophilic properties of di(organo)cuprates enable these compounds to react with organic substates to form copper(III) complexes which undergo reductive elimination may also be used to account for their reactions with propargylic esters and $\alpha\beta$-unsaturated carbonyl compounds to give allenes and 1,4-addition products respectively.

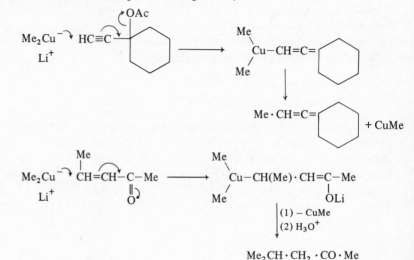

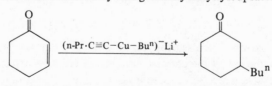

$$Me_2CH \cdot CH_2 \cdot CO \cdot Me$$

In all the reactions of di(organo)cuprates ($R_2Cu^-Li^+$) mentioned above only one of the groups (R) appears in the final organic cross-coupled product. This can represent a severe practical disadvantage if this group has been prepared by a multi-step synthesis. This disadvantage can be overcome by using mixed organocuprates, because the tendency for a group to be transferred from copper increases in the order cyanide < halide < alkynyl < aryl < alkenyl < alkyl. The selective transfer of an alkyl group for example, can therefore be achieved by using an alkylalkynylcuprate.

The chemistry of organocuprates, and the possibility that copper(III) complexes are formed during reactions of these compounds, has been discussed by Normant (1972), Posner (1972), and Jukes (1974).

5.3. Insertion

In inorganic and organometallic chemistry the term 'insertion' is used to denote the process in which an unsaturated molecule (L′) is inserted into the bond between a metal (M) and one of its ligands (L) so that a new ligand (L′−L) is created.

$$M-L + L' \rightarrow M-L'-L \qquad [5.4]$$

Examples of insertions are known in which the ligand (L) is hydrogen or a halogen, or a more complex group bonded to the metal through carbon, oxygen, nitrogen, or another metal.

All insertions are, of course, specific examples of the general process of addition, i.e. equation 5.4 represents the addition of M−L to the species L′ and, at the present time, almost all the known insertions are additions of either the α- or β-type. The latter class includes insertions of alkenes, alkynes, nitriles and sulphur dioxide, while the former includes insertions of isocyanides, carbenes and carbon monoxide.

$$M-L + \bar{C} \equiv O^{+} \rightarrow M-\underset{\underset{O}{\|}}{C}-L \qquad \text{α-addition}$$

$$M-L + H_2C{=}CH_2 \rightarrow M-CH_2-CH_2-L \qquad \text{β-addition}$$

Although there are many reactions of compounds of main-group metals that are examples of insertion and that proceed through intermediate metal complexes, e.g. the addition of Grignard reagents to carbonyl compounds (see §4.8), in this chapter the discussion is largely confined to insertions of unsaturated molecules into the bond between a transition metal and either carbon or hydrogen.

Of the transition metals whose compounds are known to participate in insertions, those with a d^6 or d^8 configuration feature very strongly, but it has been suggested that this is probably only because complexes of these metals are usually fairly stable and easy to work with, and hence have been the most extensively studied. The early work on insertions with transition-metal compounds have been discussed in detail by Candlin, Taylor and Thompson (1968).

Most insertions of interest to the organic chemist are of alkenes, alkynes and carbon monoxide into metal–hydrogen or metal–carbon bonds. For convenience, the insertion of alkenes and alkynes and the insertion of carbon monoxide are discussed in separate sections of this chapter (§ §5.4 and 5.7 respectively), and these sections are followed by discussions on synthetic applications of the two types of insertion.

5.4. Insertion of alkenes and alkynes into metal–hydrogen and metal–carbon bonds

The insertion of an alkene into a metal–hydrogen bond, i.e. the addition of a metal hydride across a carbon–carbon double bond, is the most frequently encountered type of insertion, and is exhibited by hydrides of a number of transition metals and most of the metals in Groups IA, IIA, IIIB and IVB. This type of insertion is now accepted as an essential step in a number of transition-metal catalysed processes including the hydrogenation (§5.5), isomerisation (§5.6) and hydroformylation (p. 283) of alkenes. The insertions which are discussed in this section are largely those that are relevant to these catalysed processes.

Theoretically all insertions of alkenes into M—H bonds are reversible, although not all of them are in practice. In transition-metal chemistry the reversible formation of metal alkyls from metal hydrides and alkenes has been demonstrated by the reactions of ethylene with complexes such as $RuClH(Ph_3P)_3$, $RhCl_2H(Ph_3P)_2$ and $PtClH(Et_3P)_2$. The first complex, for example, is converted in chloroform into the ethyl derivative by the action of ethylene under pressure, but the hydride is reformed when the pressure is released.

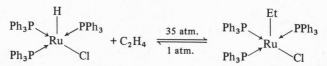

In connection with the reversibility of metal alkyl formation it must be mentioned that the thermal instability of many metal alkyls is due to the ease with which they are converted by a β-elimination into a metal hydride and alkene. If this elimination is prevented, for example by the introduction of β-substituents (see (XXVII) and (XXVIII)), then thermal stability is often considerably enhanced.

$$M(CH_2 \cdot CMe_3)_4 \qquad Cr(CH_2 \cdot CPh_3)_3$$
$$\text{(XXVII)} \qquad\qquad \text{(XXVIII)}$$

M = Ti, Zr, Hf

The insertion of an alkene into the metal–hydrogen bond of the hydride of a main-group element can proceed by an ionic, radical, or concerted mechanism; the three types are exemplified by the reactions of simple alkenes with NaH, Me_3PbH and B_2H_6 respectively. Although the same types of mechanism can be visualised for insertions with transition-metal hydrides, a high proportion of the insertions that have been studied so far are believed to involve the formation of an unstable alkene/hydride complex which re-arranges to give the metal alkyl by the hydride ligand migrating on to the co-ordinated alkene.

$$M-H + \ \overset{\diagdown}{\underset{\diagup}{C}}=\overset{\diagup}{\underset{\diagdown}{C}} \ \rightleftharpoons \ M \longleftarrow \overset{H}{\underset{\underset{\diagup\diagdown}{C}}{\overset{\overset{\diagdown\diagup}{C}}{||}}} \ \rightleftharpoons \ M-\overset{|}{\underset{|}{C}}-\overset{|}{\underset{|}{C}}-H \qquad [5.5]$$

Up to now, this rearrangement – which can be visualised as intramolecular nucleophilic attack by the hydride ligand on the co-ordinated alkene – has been directly observed with very few systems, largely on account of the high instability of metal complexes that contain a hydride and an alkene ligand in the *cis*-position of the co-ordination sphere. The most convincing example of the rearrangement has been provided by the molybdenum complex (XXIX) obtained by protonation of $Mo(C_2H_4)_2(diphos)_2$ with trifluoroacetic acid. Nuclear magnetic resonance spectroscopy shows that at 7 °C this complex rapidly and reversibly rearranges to the ethyl derivative (XXX).

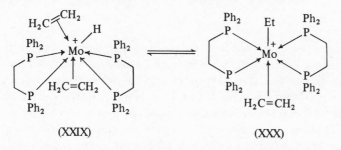

(XXIX)   (XXX)

Additional support for the reversibility of the rearrangement shown in equation 5.5 has also been provided by the thermal decomposition (180 °C) of the platinum alkyl (XXXIII) which affords the hydride (XXXI) and ethylene. Deuterium labelling experiments show that the hydride ligand in (XXXI) originates from both CH_2 and CH_3 groups of the ethyl group in the starting complex. Furthermore, the two carbon atoms of the ethyl group become equivalent before the ethylene is eliminated. This means that the loss of ethylene from the intermediate (XXXII) is a slow process compared with the formation of this intermediate from the ethyl derivative.

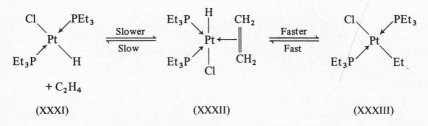

(XXXI)   (XXXII)   (XXXIII)

The conversion of the hydride (XXXI) back into the ethyl derivative by the action of ethylene requires fairly drastic conditions (18 hr/80 °C/192 atm), and necessarily involves an initial increase in the co-ordination number of the platinum during the step in which the square-planar hydride is converted into the ethylene complex (XXXII). This temporary increase in the co-ordination number of the metal occurs during reactions of alkenes with certain other metal hydrides, and, as expected, these reactions are inhibited by ligands that can successfully compete with the alkene for the required co-ordination site on the metal. The reaction between ethylene and the rhodium(III) complex (XXXIV; L = PPh_3), for example, is inhibited by triphenylphosphine which can occupy the sixth co-ordination position on the metal and thus prevent the ethylene from doing so.

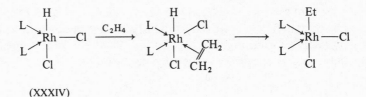

(XXXIV)

With some insertions, however, the reaction is initiated by one of the ligands dissociating from the metal to leave a vacant (or weakly solvated) co-ordination position which is then filled by the alkene. Insertions of alkenes into the metal–hydrogen bond of $CoH(CO)_4$, for example, involve the initial formation of the co-ordinatively unsaturated complex $CoH(CO)_3$.

The second step of equation 5.5, i.e. the rearrangement of the alkene/metal hydride complex, requires a *cis*-arrangement of the alkene and hydride ligands about the metal and necessarily results in an overall *syn*-addition of the metal hydride to the carbon–carbon double bond. This is in accord with the results which have been obtained from systems that are amenable to stereochemical studies.

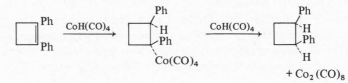

With most acyclic alkenes it is difficult to draw any conclusion concerning the stereochemistry of the reaction, for as stated earlier the rearrangement of the alkene complex into the metal alkyl is usually a reversible process, and this allows the alkene to be readily converted into geometrical and positional isomers (see §5.6), thus resulting in the formation of mixtures of stereoisomeric insertion products. This reversibility also results in the product of the insertion of an unsymmetrical alkene being controlled by thermodynamic as well as by kinetic factors, and hence it is often dependent upon reaction conditions as well as upon the actual metal hydride and alkene that are used. Thus, only the straight-chain platinum alkyl (XXXV; $R = n\text{-}C_6H_{13}$) appears to be formed when the hydride (XXXI) is heated with 1-octene, but the reaction of $CoH(CO)_4$ and propene in the gas phase gives largely ($\sim$70 per cent) the branched chain product (XXXVI) as the initial product.

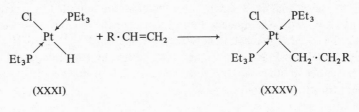

(XXXI)   (XXXV)

$$CoH(CO)_4 + Me \cdot CH=CH_2 \rightarrow Me_2CH \cdot Co(CO)_4 + n\text{-}Pr \cdot Co(CO)_4$$

(XXXVI)

One very important factor which controls the relative stabilities of the metal alkyls that can be produced is the degree of steric hindrance that arises between the ligands already present on the metal and the newly created alkyl chain. Branching at the α-position of the alkyl chain is therefore generally undesirable, and accordingly *sec*-alkyl-iridium(III) complexes readily isomerise when heated in solution to give the corresponding n-alkyl complex *via* the alkene–metal hydride intermediate.

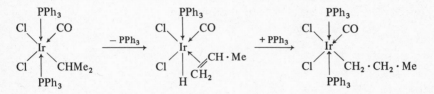

In connection with the formation of the cobalt alkyl (XXXVI), it should be noted that insertions into the metal–hydrogen bond of metal hydrido-carbonyls generally afford reactive metal alkyls which undergo further transformations *in situ*. Insertions with cobalt hydridotetracarbonyl, for example, are invariably followed by a rapid alkyl migration to give the metal–acyl derivative (see p. 269). Stable metal alkyls have only been obtained from insertions in which the alkene bears strongly electron-withdrawing substituents, or in which the metal alkyl is stabilised by chelation.

$$(CO)_5Re \cdot CF_2 \cdot CHFCl \xleftarrow{F_2C=CFCl} (CO)_5ReH \xrightarrow{} (CO)_4Re$$

The stability of the alkene–metal hydride complex which is the reaction intermediate in the insertions described so far depends upon the nature of the alkene and the metal hydride. Indeed, in an extreme case the stability may be so low that the complex represents no more than a shallow well in the energy profile for the formation of the transition state (XXXVII) and the addition could then justifiably be considered as a 1,2-concerted process.

(XXXVII)

On the other hand there are examples in which the alkene is so strongly bonded to the metal hydride that the complex does not rearrange to give the metal alkyl. In the majority of these examples the structures of the alkene and the metal hydride are such that it is more correct to regard the complex between the two reactants as a three-membered metalocycle, rather than as a π-complex; i.e. the formation of the complex constitutes an oxidative addition (see p. 223). Thus, tetracyanoethylene reacts with $Pt(CN)H(Et_3P)_2$ and $IrH(CO)(Ph_3P)_3$ to give the 'alkene' complexes (XXXVIII; X = CN) and (XL) respectively. When $PtHX(Et_3P)_2$ (X = Cl and Br) is used the initial products (XXXVIII) undergo a rapid reductive elimination of HX to give the platinum(II) derivative (XXXIX), rather than rearrange to give a σ-bonded platinum(II)–tetracyanoethyl compound.

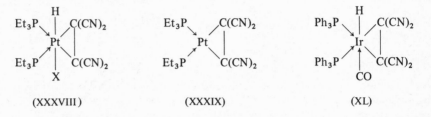

(XXXVIII)   (XXXIX)   (XL)

One other alkene–metal hydride complex whose stability is worthy of note is the salt (XLI) which can be stored indefinitely at 0 °C. Presumably this stability is due to the alkene and hydride ligands having a *trans*-arrangement about the metal, rather than the *cis* one that is necessary for migration of the hydride ligand on to the co-ordinated alkene.

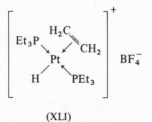

(XLI)

The insertion of an alkene into the metal–carbon bond of a transition-metal alkyl is of interest in that this particular type of insertion is widely believed to constitute the chain-extension step in the Ziegler polymerisation of 1-alkenes. In this low-temperature process a catalyst is prepared by alkylation of a compound of titanium (or another early transition-metal) by an organometallic of a main-group metal, e.g. an aluminium trialkyl. It is generally accepted that the active species of the catalyst is a co-ordinatively unsaturated titanium alkyl (XLII) whose vacant co-ordination position becomes occupied by a molecule of the alkene. Migration of the alkyl residue from the metal on to the co-ordinated alkene affords another co-ordinatively unsaturated complex (XLIII) whose vacant co-ordination position is again occupied by a molecule of alkene. Migration of the alkyl residue is followed by co-ordination of a further molecule of alkene, and this sequence of events is repeated again and again until ultimately the titanium residue is attached to a linear, polymeric alkyl chain. Finally, the metal becomes detached from this chain, possibly by the elimination of a Ti–H species.

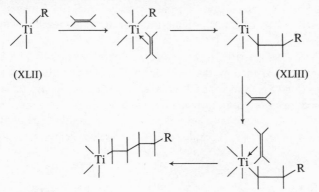

(XLII)   (XLIII)

In spite of the very wide acceptance of this mechanism for polymerisation there are no known examples of the stoichiometric insertion of a 1-alkene into a metal–carbon bond, and stable 1 : 1 insertion products have only been obtained using alkenes bearing strong electron-withdrawing substituents, as in the following examples.

$$\eta\text{-}C_5H_5 \cdot Fe(CO)_2 \cdot CH_2Ph \xrightarrow{(CN)_2C:C(CN)_2} \eta\text{-}C_5H_5 \cdot Fe(CO)_2 \cdot C(CN)_2 \cdot C(CN)_2 \cdot CH_2Ph$$

$$Mn(CO)_5Me \xrightarrow{C_2F_4} Mn(CO)_5 \cdot CF_2 \cdot CF_2 \cdot Me$$

However, there is evidence that some of these stable insertion products are formed *via* a complex in which the alkene is bonded to the metal, i.e. of the type suggested in the polymerisation mechanism. Thus in the reaction between tetrafluoroethylene and the methylplatinum(II) compounds (XLIV; L = a tertiary phosphine or arsine), intermediate complexes of type (XLV) have been isolated, and changes in the ligands (L) that would be expected to increase the affinity of the metal for the alkene, e.g. $AsMe_2Ph > PMe_2Ph > PPh_3$, increase the reactivity of the starting complex.

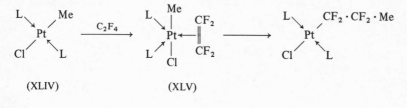

(XLIV)   (XLV)

In contrast, the few simple alkene/alkyl complexes which have been iso-
lated, e.g. $[(\eta\text{-}C_5H_5)_2$ Mo(ethylene)Me$]^+$ (PF$_6$)$^-$, have been found to be stable
with respect to insertion of the alkene into the metal-alkyl bond. This stability
casts doubts on the validity of the polymerisation mechanism outlined above,
and very recently an alternative mechanism which involves metal carbene
complexes and four-membered metallocycles and which is very similar to that
accepted for alkene metathesis (see p. 97) has been suggested. The suggestion
is supported by the observation that there is a close identity between those
catalysts used in Ziegler–Natta polymerisations and those used in alkene
metathesis.

Examples of the insertion of a conjugated double bond into a metal-
carbon bond are provided by the reactions of 1,3-dienes with π-allyl palladium
compounds to give complexes of 1,6-dienes. N.m.r. studies have shown that
some of these reactions definitely involve co-ordination of the diene with the
σ-allyl form of the palladium complex, as in the reaction between buta-1,3-
diene and the mixed π-allyl complex (XLVI; X = acetylacetonate).

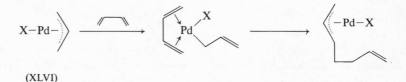

(XLVI)

Finally it should be noted that hydride complexes of several transition
metals such as Mn(I), Re(I), Fe(II), Mo(IV) and Zr(IV) undergo insertions
with alkynes to give the expected σ-alkenylmetal derivative. With $(\eta\text{-}C_5H_5)_2$
ZrHCl and dialkyl-substituted alkynes the insertions are stereospecific (*syn*-
addition) and give the product in which the metal is in the less hindered
position. As the metal can be readily replaced by halogen, the overall process
constitutes a stereospecific synthesis of alkenyl halides.

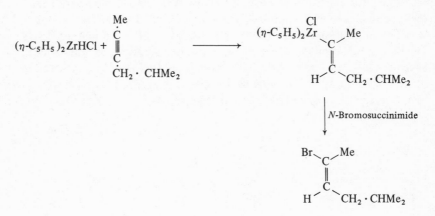

With hydrides of most other metals only alkynes with electron-withdrawing substituents afford stable products, and the stereochemistry of the insertion varies from one system to another.

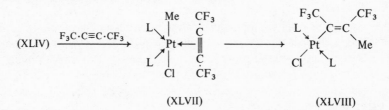

However, there is evidence that those insertions that afford the product of *syn*-addition involve an intermediate π-complex between the alkyne and the metal hydride. Support for an analogous type of intermediate in the insertion of alkynes into metal-*carbon* bonds has been provided by the reaction between the platinum(II) complex (XLIV; L = AsMe$_2$Ph) and hexafluorobut-2-yne. The initial product is the π-complex (XLVII), but in solution this rearranges to give the σ-alkenyl derivative (XLVIII).

Many examples of the insertion of an alkyne into the metal–carbon bond of a metal alkyl or a metal alkenyl are known, but of greatest interest are the insertions involving metal alkynyls. This particular type of insertion is believed to be a critical step in the transition-metal catalysed oligomerisation of alkynes, and a specific example of such an insertion is provided by the conversion of the palladium alkynyls (XLIX; X = Br or Cl, R = Et or Bu) into the corresponding alkenyl derivatives (L).

5.5. Homogeneous hydrogenation

One important result of the upsurge of activity in organometallic chemistry during the nineteen fifties and sixties was the discovery of many complexes, mainly of Group VIII metals, that could catalyse the hydrogenation of unsaturated organic compounds under homogeneous conditions. Although several examples of homogeneous hydrogenation were known prior to this period, the catalytic hydrogenation of organic compounds was always carried out using finely-divided transition metals, often on an inert support such as charcoal, calcium carbonate or barium sulphate. From the preparative point of view such heterogeneous hydrogenations were effective and convenient, and of course are still widely used, but they suffer from the limitation that few reliable conclusions concerning their mechanism of action can be drawn. This is because of the difficulty of observing the chemical and physical properties of the species chemisorbed on the catalyst, and the fact that a wide variety of reactive centres are invariably present on the surface of any particular catalyst. Related to this last point is the main disadvantage associated with the preparative use of the traditional insoluble catalysts, i.e. their activity often varies from batch to batch and is dependent upon a number of variables such as age and adventitious poisoning. The discovery of stable, soluble complexes that could function as hydrogenation catalysts and that could be obtained in a reproducible state of purity and activity enabled the mechanism of catalytic hydrogenation to be studied by a variety of experimental techniques which were inapplicable in the studies with insoluble catalysts. In many systems the properties of the chemical species that resulted from the interaction of the catalyst with hydrogen and/or the organic substrate could be accurately determined, while in some systems these species could be isolated as stable entities. As a result, a substantial amount of reliable information is now available concerning the various types of mechanism by which the soluble catalysts function. Unfortunately, however, considerable caution has to be applied when this information is extrapolated to insoluble catalysts because of their greater physical and chemical complexity.

An additional advantage of using soluble metal complexes as hydrogenation catalysts is that one can study the changes in reactivity that result from systematically modifying the chemical structure of the complex, for example by changing the ligands or altering the oxidation state of the metal. From such studies a number of metal complexes have been found to exhibit a high degree of selectivity, i.e. they catalyse the hydrogenation of only a limited range of structural types of organic compound. Examples of such complexes are given in this section, following a short discussion on the mechanisms by which hydrogenations can be catalysed by transition-metal complexes. The fact that most of this discussion refers to the hydrogenation of alkenes is a reflection of the frequency with which this type of compound has been used in hydrogenation studies. For more comprehensive discussions on homogeneous hydrogenation the reader is referred to the books by James (1973) and McQuillin (1976), and the review by Birch and Williamson (1976).

All catalytic hydrogenations are necessarily composed of a continuous cycle of reactions, the overall result of which is the addition of molecular hydrogen to a multiple bond. Several distinct types of reaction cycle have been identified, and an essential step in all of them is the interaction between a metal complex and molecular hydrogen to form either a mono- or di-hydride species. Although the precise stage at which this step occurs is dependent upon the type of reaction cycle, there currently appear to be only three general processes by which hydride formation takes place. In two of these processes hydride formation is associated with an increase in the formal oxidation state of the metal by one and by two units, while in the third the metal remains in its original oxidation state. An example of the last type is the reaction of dihydrogen with tris(triphenylphosphine)ruthenium(II) chloride to give a ruthenium(II) hydride.

$$RuCl_2(PPh_3)_3 + H_2 \rightarrow RuHCl(PPh_3)_3 + HCl$$

It can be seen that this type of process is in effect a substitution in that an anionic ligand is replaced by a hydride anion derived from the dihydrogen.

An example of hydride formation in which the metal increases its oxidation state by one unit is provided by the reaction of dihydrogen with freshly prepared solutions of the pentacyanocobaltate(II) anion to give the corresponding hydridocobaltate(III) species.

$$2[Co(CN)_5]^{3-} + H_2 \rightarrow 2[CoH(CN)_5]^{3-}$$

Largely on account of the complex nature of solutions which contain cyanide and cobalt(II) ions, the pentacyanocobaltate(II) anion is probably the most widely studied of the homogeneous hydrogenation catalysts, but for the organic chemist its main interest is its ability selectively to reduce conjugated double bonds.

The final type of reaction in which a metal hydride can be formed from the interaction between dihydrogen and a metal complex is actually a specific example of the general process of oxidative addition, and is best exemplified by the rhodium(I) complex, $RhCl(PPh_3)_3$, which was first reported by Wilkinson's group at Imperial College in 1965 and which so far has proved to be the most useful of the catalysts available for homogeneous hydrogenations.

$$RhCl(PPh_3)_3 + H_2 \rightarrow RhH_2Cl(PPh_3)_3$$

In the oxidative addition of dihydrogen the metal can be regarded as a nucleophile which causes fission of the H_2 molecule by transferring electron density into the H—H antibonding orbital. In agreement with this concept the rate of reaction and the stability of the resultant dihydride relative to the starting materials are both increased by factors that increase the nucleophilicity of the metal, e.g. a decrease of either the oxidation state of the metal or of the π-accepting properties of the ligands. Thus with the rhodium(I) complex mentioned above the rate of reaction is increased by electron-donating substituents in the *para*-positions of the phenyl groups but decreased by electron-withdrawing ones, and dihydride formation does not occur at all under mild conditions when one of the triphenylphosphine ligands is replaced by the stronger π-acceptor, carbon monoxide.

The stability and reactivity of the hydride formed by any of the three processes described above are in fact very important, in that the hydride must be stable with respect to reduction to the metal – hence the usual presence of stabilising ligands such as CO, CN, or PPh_3 – but on the other hand it must not be so stable and unreactive that it cannot participate in the next stage of the reaction cycle. This stage usually involves the conversion of the metal hydride into a metal alkyl, either by an intermolecular reaction with the alkene or by an intramolecular one which involves a hydride–alkene complex. The latter possibility, which will be considered first of all, can be illustrated by reference to the dihydride, $RhH_2Cl(PPh_3)_3$, formed from Wilkinson's catalyst. In solution one of the triphenylphosphine ligands reversibly dissociates from the dihydride to leave a co-ordinatively unsaturated species, the vacant co-ordination site of which is probably occupied by a loosely-bound solvent molecule as shown in (LI). Displacement of this solvent molecule by the alkene affords a dihydride–alkene complex (LII) which can rearrange to a metal alkyl by one of the hydride ligands migrating on to the co-ordinated alkene. A final reductive elimination of the alkane regenerates a rhodium(I) complex (possibly (LIII)) which with dihydrogen can reform the co-ordinatively unsaturated dihydride (LI); the cycle of reactions can then be repeated.

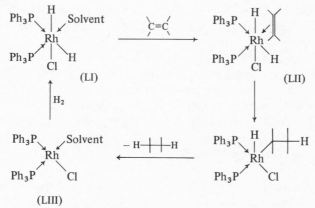

With Wilkinson's catalyst it has definitely been established that co-ordination of the alkene *via* a co-ordinately unsaturated species ((LI) in this case)

follows dihydride formation rather than precedes it. In hydrogenations with the so-called Vaska's compound, $IrCl(CO)(PPh_3)_2$ (named after its discoverer, 1962), there is kinetic evidence to suggest that the order of these two events is reversed, and the formation of $IrCl(CO)(PPh_3)(alkene)$ precedes the oxidative addition of dihydrogen. With both catalysts, however, the addition of dihydrogen occurs before the stage in which a hydride ligand migrates on to the co-ordinated alkene to form a metal alkyl. This is not always the situation, as is illustrated by the reaction cycles associated with some catalysts, e.g. $RuHCl(PPh_3)_3$, $RhH(CO)(PPh_3)_3$ and $CoH(CO)_4$, all of which necessarily contain at least one hydride ligand. With these catalysts the reaction cycle is initiated by the formation of a co-ordinatively unsaturated species by loss of a ligand and subsequently involves co-ordination of the alkene, metal alkyl formation, and then the oxidative addition of dihydrogen. A final reductive elimination of the alkane reforms the co-ordinatively unsaturated species.

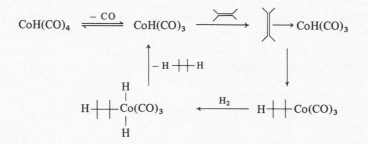

With all the catalysts that have to give rise to a co-ordinatively unsaturated species at some stage in their reaction cycle in order to allow co-ordination of the alkene, the catalytic activity is diminished by any chemical modification that reduces the ease with which the unsaturated species is formed. Taking an extreme example, although the rhodium dihydride $RhClH_2(PPh_3)_3$ is a very efficient hydrogenation catalyst the corresponding iridium complex is completely inactive because in solution all three triphenylphosphine ligands remain firmly co-ordinated with the metal. The required bis(triphenylphosphine) complex, $IrClH_2(PPh_3)_2$, can be prepared however by an indirect route, and shows the expected high activity. For the same reason, of the two trihydrides $IrH_3(PPh_3)_3$ and $IrH_3(PPh_3)_2$, the former is inactive under conditions in which the latter is an active catalyst. The activity of a homogeneous catalyst is also reduced by the addition to the hydrogenation mixture of ligands that will either inhibit the formation of the unsaturated species or strongly bind to its vacant site and thus prevent the alkene from doing so. Hydrogenations with $RhCl(PPh_3)_3$, for example, are inhibited by triphenylphosphine which represses the formation of the complex (LI), and by pyridine or benzonitrile which very effectively compete with the alkene for the solvated position on this species and consequently prevent the formation of the alkene complex

(LII). This inhibition is directly analogous – and probably mechanistically identical – to the poisoning of the traditional insoluble catalysts by potential ligands such as pyridine, sulphur compounds and iodide anions.

In the reaction cycles associated with a limited number of homogeneous hydrogenation catalysts the reaction between the intermediate metal hydride and the alkene is intermolecular, and hence does not involve an alkene–hydride complex. Consequently a co-ordinatively unsaturated complex is not an essential participant in the hydrogenation. The partial hydrogenation of conjugated dienes to mono-alkenes in the presence of pentacyanocobaltate(II) anion, for example, appears to involve an intermolecular reaction between the diene and the initially-formed hydridopentacyanocobaltate(III) anion (see above) to give a cobalt(III) alkyl, e.g. (LIV), from which the original pentacyano-cobaltate(II) anion and the mono-alkene are formed by the action of a further hydridopentacyanocobaltate(III) anion. With most dienes there is still considerable uncertainty about several aspects of the hydrogenation, e.g. whether the reaction between the diene and the hydridopentacyanocobaltate(III) anion is a 1,4- or a 1,2-addition, and to what extent the resultant cobalt(III) alkyl exists in the σ-allyl and π-allyl forms. For simplicity these possibilities are ignored in the following representation of the catalytic hydrogenation of 1,3-butadiene.

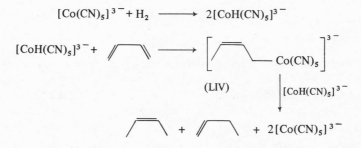

$$[Co(CN)_5]^{3-} + H_2 \longrightarrow 2[CoH(CN)_5]^{3-}$$

(LIV)

So far the discussion has concerned catalysed hydrogenations in which the transference of two hydrogen atoms derived from a molecule of hydrogen is a stepwise process which involves a metal alkyl. One or two systems are known which clearly show that this transference of hydrogen can take place by other routes. Thus, in contrast to the conjugated diene system mentioned above, the hydrogenation of alkenes catalysed by $[Co(CN)_5]^{3-}$ in which the double bond is conjugated with a $C=O$, $C=N$, or $C\equiv N$ group appears to be a radical process. In the case of the cinnamate anion, for example, the experimental evidence strongly suggests the following two-stage mechanism.

$$Ph \cdot CH = CH \cdot CO_2^- + [CoH(CN)_5]^{3-} \rightarrow Ph \cdot CH_2 - \dot{C}H \cdot CO_2^- + [Co(CN)_5]^{3-}$$

$$Ph \cdot CH_2 - \dot{C}H \cdot CO_2^- + [CoH(CN)_5]^{3-} \rightarrow Ph \cdot CH_2 - CH_2 \cdot CO_2^- + [Co(CN)_5]^{3-}$$

Metal alkyls do not play an active part in this hydrogenation, even though they are formed during the reaction. Similarly, metal alkyls do not appear to be involved in those partial hydrogenations of conjugated dienes that are catalysed by complexes of the type $M(arene)(CO)_3$ where $M = Cr, Mo, or W$. Selectivity, kinetic and deuteriation studies indicate that a 1,4-addition is the predominant route for the hydrogenation.

$$M(arene)(CO)_3 \rightleftharpoons M(CO)_3 + arene$$

$$M(CO)_3 + H_2 \rightleftharpoons MH_2(CO)_3$$

Although it would be inappropriate to discuss all the experimental results that have led to various mechanisms being suggested for hydrogenations with the different types of catalyst, it should be mentioned that in the cases where they have been obtained the stereochemical results are as predicted. For example, the formation of *cis*-3-hexene from 2,4-hexadiene and the formation of a 1 : 1 mixture of the *threo*- and *erythro*-forms of $HO_2C . CHD . C(Me)D . CO_2H$ from citraconic acid when $Cr(methyl benzoate)(CO)_3$ and $[CoD(CN)_5]^{3-}$ respectively are used as the hydrogenation catalysts are both consistent with the 1,4-addition and the radical mechanisms which have been suggested for hydrogenations with these catalysts. With catalysts of the Wilkinson type, overall *syn*-addition of H—H to the carbon–carbon double bond was predicted, for the formation of the intermediate metal alkyl from the mixed alkene–hydride complex is consistent with *syn*-addition of M—H to C=C, and the subsequent reductive elimination of alkane from the H—M–alkyl complex should take place with retention of configuration at carbon. In agreement with this prediction the addition of D_2 to maleic and fumaric acids with $RhCl(PPh_3)_3$ as catalyst affords *meso*- and ($\pm$)-1,2-dideuteriosuccinic acids respectively.

Included in the stereochemical studies are a number of investigations that show that chiral catalysts can be used for carrying out asymmetric hydrogenations (see review by Morrison, Masler and Neuberg, 1976). Elements of chirality have been introduced into homogeneous catalysts in several ways including the use of phosphine ligands, e.g. (LV) and (LVI), that bear chiral substituents (see p. 254).

Hydrogenation of 2-acetamidocinnamic acid (LVII) in the presence of a catalyst generated *in situ* by displacement of the cycloalkene ligands from $[RhCl(cyclo\text{-}octene)_2]_2$ by the diphosphine (LV), for example, affords *N*-acetyl-(D)-phenylalanine (LVIII) with an optical yield of 72 per cent.

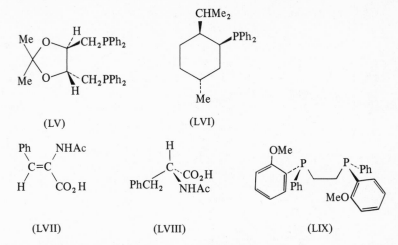

(LV)   (LVI)

(LVII)   (LVIII)   (LIX)

Exceptionally high optical yields (> 90 per cent) can be obtained in the homogeneous hydrogenation of α-acylaminoacrylic acids such as (LVII) by use of optically active rhodium–phosphine complexes in which the chirality resides on the phosphorus atom, as in the diphosphine (LIX). The use of these complexes therefore provides a convenient route to almost optically pure α-amino acids.

Because of the specific way in which the intermediate transition-metal hydride chemically interacts with the unsaturated organic substrate during a catalytic homogeneous hydrogenation, there is usually a very wide variation in the catalytic activity that a particular transition-metal complex exhibits with respect to the catalysed hydrogenation of different types of organic compounds. Tris(triphenylphosphine)chlororhodium(I), $RhCl(PPh_3)_3$, is the catalyst that to date has been studied most extensively from this viewpoint. At room temperature and 1 atm pressure of dihydrogen this complex efficiently catalyses the hydrogenation of alkenes and alkynes, but is ineffective in the reduction of keto, hydroxy, cyano, nitro, chloro, ether, and carboxylic acid and ester groups to name but a few. With alkenes the rates of hydrogenation are dependent on steric factors, and are highest and lowest with monosubstituted and tetrasubstituted alkenes respectively. Indeed, with tetrasubstituted alkenes, and to a lesser extent with trisubstituted, higher temperatures and pressures usually have to be employed. The steric factors are believed to come into play during the formation of the dihydride–alkene complex (LII) from the co-ordinatively unsaturated precursor (LI) (see p. 250). The effect of substitution on the rates of hydrogenation of alkenes, together with the inertness of various functional groups, has allowed the selective reduction of many multi-functional compounds, five of which are indicated by structures (LX)–(LXIV) together with the double bonds which have been reduced.

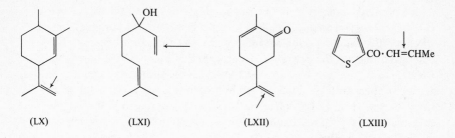

(LX)   (LXI)   (LXII)   (LXIII)

The successful reduction of the double bond in the side-chain of the thiophen (LXIII) illustrates one of the advantages of RhCl(PPh$_3$)$_3$ as a catalyst, namely that the inhibition of catalytic activity by sulphur compounds is usually very small. Another advantage is that with unsaturated six-membered ring carbo-cycles very little disproportionation occurs, and the catalytic hydrogenation of (LXIV), for example, affords only 4 per cent of the aromatic (LXV).

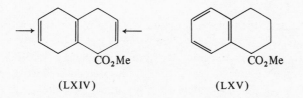

(LXIV)   (LXV)

As stated above, ketones are not reduced by the RhCl(PPh$_3$)$_3$/H$_2$ system. Sterically unhindered keto groups are reduced, however, by a solution of the rhodium complex and trimethyl phosphite in hot 2-propanol, which also functions as the hydrogen donor. With six-membered ring ketones the reduc-tion is stereoselective and gives predominantly (> 95 per cent) the axial alcohol. In both the 5-α and 5-β series of steroids, for example, 2- and 3-keto groups are reduced to the 2-β and 3-α alcohol respectively.

Another rhodium(I) complex that is of use in selective catalytic hydrogena-tions is the hydride, RhH(CO)(PPh$_3$)$_3$. This complex catalyses the hydrogenation of non-conjugated terminal alkenes, but not the hydrogenation of conjugated dienes, or of cyclic and acylic di-, tri- and tetra-substituted alkenes.

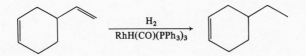

Although this complex shows higher selectivity than RhCl(PPh$_3$)$_3$, it is less effective as a catalyst. This is not true however of RuHCl(PPh$_3$)$_3$, which is more efficient than RhCl(PPh$_3$)$_3$, and which also selectively catalyses the hydrogenation of non-conjugated terminal alkenes.

In complete contrast to the three complexes described above, the penta-cyanocobaltate(II) anion, $[Co(CN)_5]^{3-}$, can catalyse the hydrogenation of alkenes only when the double bond is part of a conjugated system. With conjugated dienes, the product of hydrogenation is dependent upon the CN^-/Co ratio, and can therefore be altered by the addition of cyanide anion to the reaction mixture. Thus, in the hydrogenation of 1,3-pentadiene the three alkenes (LXVI), (LXVII) and (LXVIII), are obtained in the yields shown when CN^-/Co is 7.0, but in yields of 2, 0, and 98 per cent respectively when the ratio is decreased to 5.0.

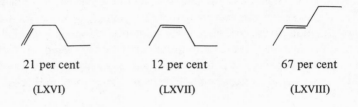

21 per cent	12 per cent	67 per cent
(LXVI)	(LXVII)	(LXVIII)

This dependence of the product upon the CN^-/Co ratio is thought to arise from the effect of CN^- on the equilibrium between the σ- and π-allyl forms of the cobalt(III) alkyl formed during the hydrogenation (see p. 252).

A highly selective method for converting conjugated dienes into monoenes, and one that does not afford the mixtures of products associated with the use of $[Co(CN)_5]^{3-}$, employs ultraviolet light to generate a thermally active hydrogenation catalyst from hexacarbonylchromium(0). These photo-assisted hydrogenations involve a 1,4-addition of dihydrogen and only proceed when the diene can easily assume the *s-cis*-conformation. Consequently the *trans-trans*-isomer of 2,4-hexadiene affords only *cis*-3-hexene, while the *cis–cis*-isomer is completely unaffected because of the high steric strain present in the required conformation.

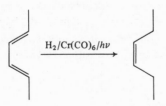

A high degree of stereoselectivity is also exhibited by the complexes $Rh(\eta\text{-}C_5Me_5)Cl_2$ and $Co(\eta\text{-}C_3H_5)[P(OMe)_3]_3$, both of which catalyse the hydrogenation of a number of unsaturated systems including arenes. These are completely reduced to the corresponding cyclohexane, and an interesting feature of the reductions is that all six hydrogen atoms are added to the same side of the ring. The hydrogenation of 1,3,5-trimethylbenzene therefore gives *cis–cis*-1,3,5-trimethylcyclohexane, while benzene and dideuterium give *all-cis-*^{2}H$_6$-cyclohexane.

The norbornadiene (NBD) complex [Rh(NBD)(PPhMe$_2$)$_3$]$^+$ PF$_6^-$ is a selective catalyst for the hydrogenation of alkynes, and with disubstituted systems affords exclusively the *cis*-alkene.

$$Me_2C(OH) \cdot C \equiv C \cdot Me \xrightarrow{H_2} Me_2C(OH) \cdot CH \overset{cis}{=\!=\!=} CH \cdot Me$$

$$Ph \cdot C \equiv C \cdot CO_2Et \xrightarrow{H_2} Ph \cdot CH \overset{cis}{=\!=\!=} CH \cdot CO_2Et$$

Indeed, as there is virtually no further reduction to the alkane stage this catalyst is superior to the Lindlar catalyst which is traditionally used for preparing *cis*-alkenes from alkynes.

The final examples of selectivity relate to the recent application of polymer-bound transition-metal complexes to catalytic hydrogenation. One of the disadvantages of all the catalysts described so far in this section is encountered when the catalyst has to be separated from the reaction products at the end of the hydrogenation. By virtue of their solubility the soluble catalysts are not nearly so readily removed as the traditional insoluble ones which are simply filtered off, and problems of product contamination and loss of catalyst can therefore arise. Recently this disadvantage has been overcome by anchoring the soluble catalysts to organic polymers, thus effectively rendering them insoluble. One technique is to introduce diphenylphosphine residues into a resin (usually of the polystyrene type) and then use the resultant polymeric phosphine as a replacement ligand for one of the triphenylphosphine ligands in a soluble catalyst, for example as in RhH(CO)(PPh$_3$)$_2$(polymer–PPh$_2$). Although in physical terms there is often very little distinction between these polymeric catalysts and the traditional heterogeneous ones, it appears that the former type still react mechanistically in the same manner as their homo-geneous counterparts, and therefore tend to exhibit the same degrees of selectivity and activity. They can therefore be used not only in hydrogenations but also in other transition-metal catalysed processes such as hydroformylation and alkene polymerisation. One of their features as hydrogenation catalysts is that they show substantial selectivity on the basis of the molecular size of the compound being hydrogenated, and cyclohexene for example, is hydro-genated much faster than a Δ^2-steroid. In addition, polar alkenes and non-polar ones are hydrogenated at different rates, the relative rates being con-trolled by the polarity of the solvent. Both these effects arise from the different concentration gradients established by different reactants within the polymer matrix.

Applications of polymer-bound complexes to catalytic hydrogenation and other processes such as hydroformylation (p. 283) have been reviewed by Hartley and Vezey (1977).

5.6. Metal-catalysed isomerisation of alkenes

Alkene isomerisation in the sense of the interconversion of *cis/trans*-isomers and positional isomers, is catalysed not only by acids and bases but also by numerous transition-metal compounds, many of which are effective under mild conditions (see reviews by Orchin (1966), Bird (1967) and Davies (1967)). Amongst these transition-metal catalysts, compounds of Group VIII metals feature very strongly, and in fact the most frequently used catalysts are either complex halides of palladium, platinum, rhodium and ruthenium, or carbonyls of iron and cobalt. Metal-catalysed isomerisation is important in that not only does it have applications in organic synthesis, but it can also represent an undesirable side-reaction in transition-metal catalysed transformations in which alkenes are involved, e.g. hydroformylation. This last point is particularly important in large-scale industrial processes, of course, where the isomerisation can lead ultimately to the formation of considerable amounts of unwanted products.

Considering first of all some of the practical aspects of those alkene isomerisations that are promoted by metal complexes, the general observation can be made that the metal complexes are usually used in either catalytic or stoichiometric amounts. In the former case one (ultimately) obtains a mixture of isomeric alkenes whose relative proportions are determined by thermodynamic factors, while in the latter case the same isomeric alkenes are obtained, but generally in the form of metal complexes. The relative proportions of these complexes are also determined by thermodynamic factors, but these proportions may be completely different of course from those of the parent alkenes in the mixture obtained by use of catalytic quantities of the metal complex.

A good example of catalytic quantities of a metal complex being used to effect isomerisation is provided by the conversion of ethyl 2-octenoate into the thermodynamically-controlled mixture of all six possible isomers under the influence of 3 per cent pentacarbonyliron(0) in refluxing octane. The mixture of alkenes obtained has the composition shown, and the relative instability of the conjugated 2,3-isomer (compared with the 6,7-isomer, for example) as the result of the negative inductive effect of the ester group is noteworthy:

Position of double bond	2,3	3,4	4,5	5,6	6,7	7,8
Per cent present	18	8	21	24	28	1

This particular isomerisation could also be effected, of course, by use of strong bases (e.g. EtONa in EtOH), but these would also promote side-reactions (e.g. Michael addition) which are completely absent under the mild conditions associated with the metal carbonyl. A similar observation can be made concerning the isomerisation of 1-methoxy-2,5-cyclodienes such as (LXIX). This isomerisation can be effected either by 1 per cent $RhCl(PPh_3)_3$ in refluxing chloroform or, less conveniently, by sodamide in liquid ammonia.

Indeed, in the case of (LXIX) the metal-catalysed isomerisation affords the most stable isomer (LXXI), in contrast to the base-catalysed one which stops at the 1,3-cyclodiene (LXX) on account of the very low stability of the anion that has to be formed in the next stage of the isomerisation.

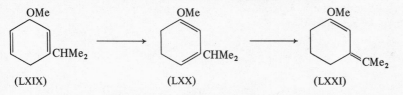

(LXIX)   (LXX)   (LXXI)

In some metal-catalysed systems the different rates at which the double bonds in the isomeric alkenes migrate allow the isolation of a mixture of alkenes whose composition is determined by kinetic rather than thermo-dynamic factors. Thus the relatively fast rate at which the double bond of a monosubstituted alkene migrates compared with that of a disubstituted alkene (particularly if the latter function is exocyclic) allows a convenient preparation of the non-conjugated isomers (LXXII; *a* and *b*), whose isomerisation to the conjugated forms (LXXIII; *a* and *b*) occurs relatively slowly.

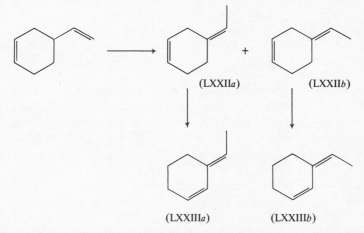

(LXXIIa)   (LXXIIb)

(LXXIIIa)   (LXXIIIb)

Similarly, the fact that the first product in the isomerisation of a 1-alkene tends to be predominantly the *cis*-form of the 2-alkene and that this iso-merises to the corresponding *trans*-form at a slower rate than it is formed, allows a convenient preparation of *cis*-2-alkenes, e.g. *cis*-isosafrole (LXXV) from safrole (LXXIV).

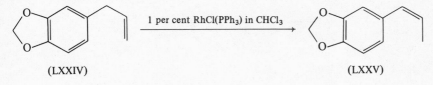

(LXXIV)   (LXXV)

1 per cent RhCl(PPh₃) in CHCl₃

Another synthetic application of metal complexes as isomerisation catalysts is in the use of an allyl group as a protecting group for alcohols. This group can be introduced by the action of allyl bromide and sodium hydroxide (or sodium hydride) in benzene, and readily removed by refluxing with RhCl $(PPh_3)_3$ in aqueous ethanol followed by mild acid hydrolysis (pH ~2) of the resultant 1-propenyl ether.

$$R \cdot OH \xrightarrow{BrCH_2 \cdot CH : CH_2} R \cdot O \cdot CH_2 \cdot CH : CH_2 \xrightarrow{RhCl(PPh_3)_3} R \cdot O \cdot CH : CH \cdot CH_3$$

$$\downarrow H_3O^+$$

$$R \cdot OH$$

The most frequently used complex that is employed in stoichiometric amounts for promoting alkene isomerisation is undoubtedly pentacarbonyl-iron(0). One advantage of this particular compound is that with conjugated dienes that can adopt an s-*cisoid* conformation it forms stable complexes of the general type (LXXVI) from which the iron tricarbonyl residue can be

(LXXVI)

easily removed to give the diene in a high state of purity (see p. 92). Consequently the pentacarbonyl is particularly useful for isomerising non-conjugated dienes to the conjugated variety, for (subject to the stereochemical requirement mentioned above) the latter are easily isolated from the isomerisation mixture in the form of their iron tricarbonyl complexes. Examples of such isomerisations include the conversion of cyclo-1,5-octadiene to the 1,3-isomer (LXXVII). The conversion of heteroannular steroidal dienes of general type (LXXVIII) into the thermodynamically less stable homoannular isomers (LXXIX) obviously involves the same principle (see p. 100).

With respect to the mechanism of metal-catalysed isomerisations, very little is known about the finer details of the cycle of reactions that constitutes the overall isomerisation, and which almost certainly varies from one catalyst to another. However, those isomerisations that have been examined

(LXXVII)

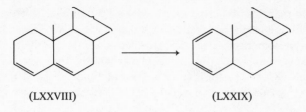

(LXXVIII)   (LXXIX)

mechanistically appear to proceed by reaction cycles that involve one of two distinct reaction sequences, both of which are initiated by the formation of a π-complex of the starting alkene. In some isomerisations this complex is formed by the addition of the alkene to a co-ordinatively unsaturated complex generated from the catalyst either by the dissociation of a ligand (as with $Fe(CO)_5$ and $IrH_3(PPh_3)_3$) or by thermal or photolytic cleavage of a poly-nuclear system (as with $Fe_3(CO)_{12}$ and $Co_2(CO)_8$).

$$IrH_3(PPh_3)_3 \xrightarrow{-PPh_3} IrH_3(PPh_3)_2 \xrightarrow{\quad} \quad \longrightarrow IrH_3(PPh_3)_2$$

$$Co_2(CO)_8 \xrightarrow{\Delta} 2Co(CO)_4 \xrightarrow{2} 2 \quad \longrightarrow Co(CO)_4$$

In what is often termed the 'π-allyl mechanism' for alkene isomerisation, and which is illustrated in figure 5.1 for the interconversion of a *trans*-2-alkene and a *trans*-3-alkene, the π-complex (LXXX) of the starting alkene rearranges to a π-allyl complex (LXXXI) by the metal undergoing a reversible oxidative addition of an allylic C—H bond. The reverse reductive elimination can afford, of course, either the initial π-complex (LXXX) of the starting alkene or the π-complex (LXXXII) of an isomeric alkene in which the double bond has been moved along the chain by one carbon–carbon unit. Displacement of this isomeric alkene from the latter complex by the starting alkene reforms the initial π-complex (LXXX), thus completing the reaction cycle. It

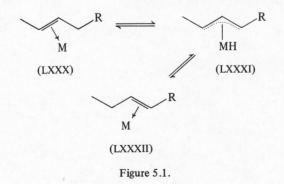

(LXXX)   (LXXXI)

(LXXXII)

Figure 5.1.

should be noted that this isomerisation involves a 1,3-hydrogen shift, and this is consistent with the results of labelling experiments in which deuteriated alkenes have been isomerised under the influence of pentacarbonyliron(0), a catalyst invariably associated with the π-allyl mechanism. Thus, the 1,4-cyclohexadiene (LXXXIII) affords (LXXXIV), and the deuteriated allyl alcohol (LXXXV) affords the propanal (LXXXVII) *via* the enol form (LXXXVI).

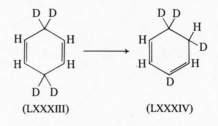

$$CH_2=CH\cdot CD_2OH \rightarrow CH_2D-CH=CDOH \rightarrow CH_2D-CH_2-CDO$$

(LXXXV)   (LXXXVI)   (LXXXVII)

As most π-allyl complexes exist in equilibrium with the corresponding σ-allyl form, the π-allyl mechanism also accounts for the interconversion of *cis-* and *trans-*isomers. This is illustrated in stereochemical terms in figure 5.2 for the *cis–trans*-equilibration of a 2-alkene. The π-allyl complex (LXXXIX), formed reversibly from the complex (LXXXVIII) of the *trans-*alkene, is in equilibrium with the σ-allyl form (XC). Free rotation about the C—C bond shown allows this species to exist in equilibrium with the π-allyl complex (XCI), which in turn is in equilibrium with the complex (XCII) of the *cis-*alkene. Both the complexes (LXXXVIII) and (XCII) are in equilibrium, of course, with the corresponding free alkene.

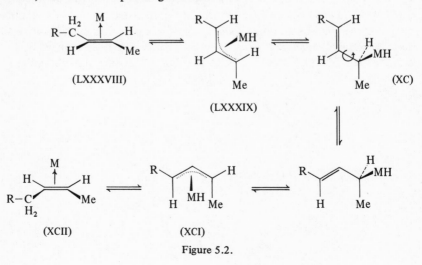

Figure 5.2.

The other main type of reaction sequence that results in either positional or *cis–trans*-isomerisation, or both of these, is initiated by the formation of a metal complex in which the metal is bonded to the starting alkene and also to a hydride ligand. This complex is represented by the structure (XCIII) in figure 5.3, which illustrates the conversion of a *trans*-2-alkene into the *cis*-2- and *trans*-3-isomers. Migration of the hydride ligand on to the co-ordinated alkene (see §5.4) gives the metal alkyl (XCIV). As this migration is reversible, the metal alkyl can proceed to the π-complex (XCV) of the *trans*-3-isomer and – after free rotation about the C—C bond shown – the π-complex (XCVI) of the *cis*-2-form. By means of similar rearrangements involving a series of metal alkyls in which the metal appears at consecutive positions along the carbon chain the π-complex (XCV) can lead to other positional and geo-metrical isomers.

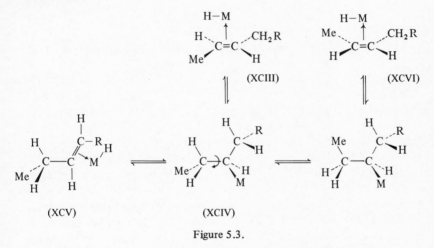

Figure 5.3.

In more general terms, this particular mechanism for isomerisation involves the reversible addition and elimination of M—H to the double bond, whose migration along the chain results in a series of 1,2-hydride shifts, in contrast to the 1,3-shifts associated with the π-allyl mechanism. These 1,2-shifts have been confirmed by the use of deuteriated alkenes in isomerisations catalysed by complexes such as $CoH(CO)_4$, $RuCl_2(PPh_3)_3$ and $RhCl(PPh_3)_3$, which favour the addition/elimination mechanism. An identical elimination/addition sequence is almost certainly operative in the reaction which occurs at room temperature between alkenes and the zirconium hydride $(\eta\text{-}C_5H_5)_2ZrHCl$. This reaction affords zirconium alkyls in which the metal is in the least hindered position of the carbon skeleton regardless of the original position of the double bond. Evidently the zirconium alkyl initially formed isomerises with great ease and ultimately forms the most stable isomer. As zirconium alkyls are cleaved by a variety of electrophilic reagents, this particular reaction provides a very useful method of functionalising non-activated positions in alkenes (figure 5.4).

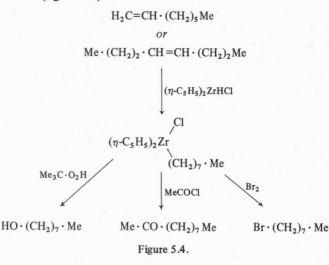

Figure 5.4.

One final point concerning the addition/elimination mechanism for the isomerisation of alkenes is that it requires the generation of a metal complex in which the metal is co-ordinated to a hydride ligand as well as to the starting alkene. Now in some isomerisations the metal complex that is used as the catalyst already contains this hydride ligand, e.g. those which use $CoH(CO)_4$, $RuHCl(PPh_3)_3$ and $RhH(PPh_3)_3$, but in others the hydride ligand becomes bonded to the metal by the catalyst being chemically modified at an early stage of the reaction. This explains why some catalysts have to be used in association with a promoter, which acts – either directly or indirectly – as the source of the hydride ligand. Examples of catalyst/promoter combinations are Li_2PdCl_4/CF_3CO_2H and $Rh(C_2H_4)_2(acac)/HCl$, where presumably the acidic components provide the necessary hydride ligand by protonating or oxidatively adding to the metal. Primary and secondary alcohols are also used as promoters (as in $RhCl_3/Me_2CH.OH$) and with these the hydride ligand is formed by hydride transfer from the α-carbon atom of an initially-formed metal alkoxide.

$$\begin{array}{c}\diagdown\\[-4pt]\diagup\end{array}CH\cdot OH + MX \; \underset{}{\overset{-\,HX}{\rightleftharpoons}} \; \begin{array}{c}\diagdown\\[-4pt]\diagup\end{array}CH-O \underset{\diagdown M}{} \; \rightleftharpoons \; \begin{array}{c}\diagdown\\[-4pt]\diagup\end{array}C=O \underset{\diagdown MH}{}$$

(XCVII)

This transfer is analogous, of course, to that involved in the rearrangement of a metal alkyl into the π-complex of an alkene and metal hydride, while its reversibility – together with the ease with which the aldehyde or ketone in the general structure (XCVII) can be displaced by other ketones – provides a mechanistic explanation for the fact that in the presence of complexes such as H_2IrCl_6 and $RhCl(PPh_3)_3$ hot 2-propanol reduces ketones to the corresponding secondary alcohol (see p. 255). The close similarity between this particular reduction and the Meerwein–Pondorf reduction (p. 214) is noteworthy. The initial formation of a metal alkoxide which then rearranges to a metal hydride and an aldehyde or ketone also explains the ability of a number of perfluorocarboxylato complexes of Group VIII metals, e.g. $Ru(O.CO.CF_3)_2$ $(CO)(PPh_3)_2$, to act as homogeneous catalysts for the dehydrogenation of primary and secondary alcohols.

5.7.   Insertion of carbon monoxide and isonitriles into metal–carbon bonds

In §1.4 it was seen that the bond between a transition metal and co-ordinated carbon monoxide can be regarded as containing two components, a metal–carbon σ-bond formed by donation of a pair of electrons on carbon into a vacant hybrid orbital of the metal and a π-bond formed by back-donation of electrons from the metal into an empty antibonding orbital of the carbon monoxide. In the case of carbon monoxide co-ordinated to a formally zero-valent metal, σ-bond formation can be represented by the resonance structure (XCVIII), but (XCIX) and (C) also take into account the formation of the π-bond.

$$\overset{-}{M}-C\equiv\overset{+}{O} \longleftrightarrow M=C=O \longleftrightarrow M=\overset{+}{C}-O^-$$

$$\text{(XCVIII)} \qquad\qquad \text{(XCIX)} \qquad\qquad \text{(C)}$$

Co-ordinated carbon monoxide can be attacked by nucleophiles at the carbon atom.

$$\overset{-}{M}-C\equiv\overset{+}{O} \longrightarrow \overset{-}{M}-C=O \longleftrightarrow M=C-O^-$$

As expected, the susceptibility to nucleophilic attack is highest when there is only a low degree of back-donation from the metal to the ligand, i.e. when the carbon monoxide is acting mainly as a donor ligand. The best example of this situation is provided by metal carbonyl cations, and a number of these species have been found to react rapidly with nucleophiles, even with alcohols under neutral conditions (see top of p. 266).

Their susceptibility to nucleophilic attack explains why the carbon monoxide ligands in cationic complexes rapidly exchange oxygen in aqueous solution.

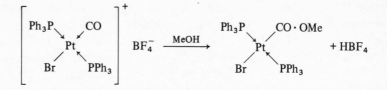

In $H_2{}^{18}O$ the cation $[Re(CO)_6]^+$, for example, is significantly enriched within minutes in contrast to most neutral metal carbonyls which are usually unreactive.

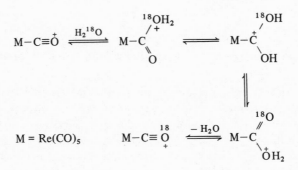

As the degree of back-bonding to the ligands is increased by an increase in the electron density on the metal, co-ordinated carbon monoxide in anionic complexes is comparatively inert to nucleophilic attack, and $[PtCl_3(CO)]^-$, for example, is much less reactive than $PtCl_2(CO)_2$. Generally, nucleophilic attack on neutral carbonyls requires the use of fairly strong nucleophiles such as amines, hydroxide anion, or Grignard and organolithium reagents. With the last type of nucleophile some carbonyls, e.g. $M(CO)_6$, M = Cr, Mo and W, afford stable, crystalline lithium salts which can be *O*-alkylated to give complexes in which the new ligand can be regarded as a carbene (see p. 96).

$$Cr(CO)_6 \xrightarrow{\text{PhLi}} \left[(CO)_5\bar{C}r-C \overset{\displaystyle Ph}{\underset{\displaystyle O}{\big/}} \right] Li^+$$

$$\Big\downarrow Et_3O^+BF_4^-$$

$$(CO)_5\bar{C}r-\overset{+}{C} \overset{\displaystyle Ph}{\underset{\displaystyle OEt}{\big/}} \quad \equiv \quad (CO)_5Cr \leftarrow : C \overset{\displaystyle Ph}{\underset{\displaystyle OEt}{\big/}}$$

The salts produced from organolithium compounds and tetracarbonyl-nickel(0) are unstable, but they can be used *in situ*, and they react with

$\alpha\beta$-unsaturated ketones, for example, to give excellent yields of 1,4-dicarbonyl compounds.

$$Ni(CO)_4 \xrightarrow{\text{n-BuLi}} [\text{n-Bu} \cdot CO \cdot \bar{N}i(CO_3)] \, Li^+$$

$$\downarrow \begin{array}{l} (1) \ Me_2C:CH\cdot CO\cdot Me \\ (2) \ H_3O^+ \end{array}$$

$$\text{n-Bu} \cdot CO \cdot CMe_2 \cdot CH_2 \cdot CO \cdot Me$$

This example of a 1,4-addition to an $\alpha\beta$-unsaturated carbonyl compound is directly analogous to the reaction that occurs with lithium diorganocuprates, (see p. 238) and the mechanisms of the two reactions are probably very similar (cf. p. 292).

$$Me_2Cu^-Li^+ \xrightarrow[\text{(2) } H_3O^+]{\text{(1) } Me\cdot CH:CH\cdot CO\cdot Me} Me_2CH \cdot CH_2 \cdot CO \cdot Me$$

When attacked by hydroxide anion, or by water in the case of reactive cationic complexes, metal carbonyls afford complexes of the type $M \cdot CO_2H$ which decarboxylate to give the metal hydride.

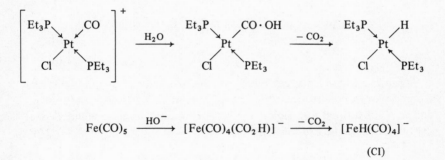

$$Fe(CO)_5 \xrightarrow{HO^-} [Fe(CO)_4(CO_2H)]^- \xrightarrow{-CO_2} [FeH(CO)_4]^-$$

$$(CI)$$

As protonation of a metal hydride (MH) normally affords molecular hydrogen and the corresponding species M^+, the formation (and subsequent decarboxylation) of complexes of the type $M.CO_2H$ from water and metal carbonyls is of interest in connection with the role played by those inorganic compounds that catalyse the formation of H_2 and CO_2 from H_2O and CO under 'water/gas' conditions.

A more specific observation is that the anionic iron hydride (CI) can reduce polarised double bonds. As this hydride can be conveniently generated from pentacarbonyliron(0) and potassium hydroxide in methanol/water (95 : 5), this mixture effectively reduces imines and $\alpha\beta$-unsaturated carbonyl compounds to the saturated derivatives; in the latter case reduction of the carbonyl group is negligible.

$$Ph \cdot CH = N \cdot Ph \longrightarrow Ph \cdot CH_2 \cdot NHPh$$

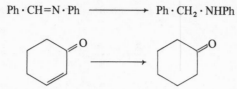

The same hydride also reduces acid chlorides to aldehydes in excellent yield under very mild conditions, thus providing an alternative to the traditional Rosenmund reduction.

$$R-CO-Cl \qquad\qquad R-CO \qquad \xrightarrow[\text{elimination}]{\text{Reductive}} \qquad R \cdot CHO$$
$$H-\bar{F}e(CO)_4 \qquad\qquad H-Fe(CO)_4 \qquad\qquad\qquad + Fe(CO)_4$$

Similar to the decarboxylation of complexes of type $M \cdot CO_2H$ is the elimination of isocyanates from the analogous carbamoyl complexes, $M \cdot CO \cdot NHR$, which are formed from metal carbonyls by nucleophilic attack with primary amines. Thus isocyanates are formed *via* an initial metal carbonyl complex when a mixture of palladium(II) chloride, carbon monoxide and a primary amine is heated at 65–85 °C.

$$PdCl_2 \xrightarrow{CO} \longrightarrow Pd \leftarrow CO \xrightarrow[-HCl]{R \cdot NH_2} \longrightarrow Pd-C-NR \longrightarrow O=C=NR$$

$$+ HCl$$
$$+ Pd(0)$$

In most cases the metal-catalysed reaction between carbon monoxide and a primary or secondary amine yields a mixture of an *N,N'*-substituted urea and an *N*-substituted formamide, the proportions of which are highly dependent upon the metal and the amine.

$$Me_2NH \xrightarrow{CO/Co_2(CO)_8} Me_2N \cdot CHO \qquad 60 \text{ per cent}$$

$$Et_2NH \xrightarrow{CO/NiI_2} Et_2N \cdot CO \cdot NEt_2 \qquad 72 \text{ per cent}$$

Although with primary amines the substituted urea could be formed from an intermediate isocyanate, with secondary amines this is not possible, and it is likely that in both cases the urea is formed by nucleophilic displacement of the metal (see p. 275) from a hydridocarbamoyl complex.

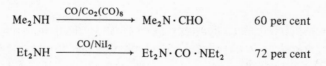

The *N*-substituted formamides are possibly formed by a reductive elimination from the same type of hydride carbamoyl complex, as illustrated by the stoichiometric reaction that occurs when pentacarbonyliron(0) is treated with secondary amines.

$$Fe(CO)_5 + R_2NH \longrightarrow (CO)_4\bar{F}e-\underset{\underset{O}{\|}}{C}-\overset{+}{N}HR_2 \longrightarrow (CO)_4 Fe \cdot \underset{\underset{H}{|}}{CO} \cdot NR_2$$

$$\Big\downarrow - R_2N \cdot CHO$$

$$Fe(CO)_4(R_2NH) \xleftarrow{R_2NH} Fe(CO)_4$$

In all the reactions described so far, it is suggested that the co-ordinated carbon monoxide reacts intermolecularly with the nucleophile. Closely related to this process is a rearrangement that is exhibited by many metal carbonyls and which is very important from the synthetic point of view. In this rearrangement a σ-bonded alkyl group or, less commonly, an aryl group, migrates from the metal to the carbon of a carbon monoxide ligand in an adjacent (*cis*) position to form a metal–acyl derivative, i.e. the overall result is that the carbon monoxide is inserted into the metal–carbon σ-bond. The migration, which is often reversible, can be viewed as intramolecular nucleophilic attack on co-ordinated carbon monoxide, and is nearly always accompanied by the co-ordination of another ligand (L) with the metal.

$$L: \quad {}^-\underset{\overset{|}{M}}{\overset{\overset{R}{\diagup}}{}}-C\equiv O^+ \quad \rightleftharpoons \quad \overset{\overset{L}{\downarrow}}{M}-\underset{\underset{O}{\diagdown\!\backslash}}{\overset{\diagup R}{C}}$$

As the reaction involves formation of the metal–ligand bond (M—L) and breakage of the metal–alkyl bond (M—R), it is possible to imagine a spectrum of mechanisms available for the reaction, each mechanism being characterised by the extent to which bond formation and bond breakage has occurred in the transition state. In one extreme mechanism formation of the M—L bond is complete before the M—R bond starts to break, while in the other extreme mechanism bond breakage is complete before the ligand begins to co-ordinate with the metal. Midway between these two possibilities is the mechanism in which the migration of the group R and formation of the M—L bond are completely synchronous. The kinetic data which have been obtained from studies of a rather limited number of reactions, e.g. [5.6] and [5.7], suggest that in polar solvents the second of the extreme mechanisms operates and the reactions are initiated by the reversible formation of a co-ordinatively unsaturated metal–acyl derivative which is then attacked by the ligand.

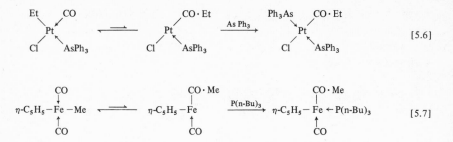

The reactions studied often show large negative entropies of activation and the rates are highest in solvents that have good co-ordinating properties. This is consistent with the possibility that in the intermediate acyl derivative a solvent molecule occupies the co-ordination position that was formerly filled by the group R.

In contrast, with non-polar solvents that have very low co-ordinating power, e.g. n-hexane, the data suggest that migration of the group R is accompanied by co-ordination of the ligand L.

The ligands L that have been used in studies of the alkyl migration reaction include halide anions, ammonia, primary amines, tertiary phosphines and phosphites, and carbon monoxide. The reversible reaction between the last compound and alkylmanganese pentacarbonyls has been examined in some detail. In agreement with the proposed mechanism it has been shown that the acetylmanganese pentacarbonyl formed by treatment of methylmanganese pentacarbonyl with ^{14}CO contains none of the label in the acetyl group, and conversely the thermal decarbonylation of the acetyl compound with a CO-labelled acetyl group affords unlabelled carbon monoxide.

$$Me \cdot {}^{*}CO \cdot Mn(CO)_5 \xrightarrow{\Delta} Me \cdot Mn({}^{*}CO)(CO)_4 + CO$$

$$Me \cdot Mn(CO)_5 + {}^{*}CO \longrightarrow Me \cdot CO \cdot Mn({}^{*}CO)(CO)_4$$

The ease with which alkylmanganese pentacarbonyls react with carbon monoxide to give the corresponding acyl derivative depends on the nature of the alkyl group, and varies in the order i-Pr $>$ Et $>$ Me $\gg$ CF$_3$, the last compound reacting immeasurably slowly. This order reflects the relative strengths of the metal–alkyl bonds in the alkyl carbonyls. The additional bond strength, which partly arises from π-bond formation between the metal and the unsaturated hydrocarbon residue, probably also explains the reluctance of Ph . Mn

(CO)$_5$ to give benzoylmanganese pentacarbonyl when treated with carbon monoxide. The same explanation accounts for the fact that although with triphenylphosphine the complex (CII; R = Me) gives the product of methyl migration (CIII), the analogous acetylide (CII; R = C≡C . Ph) undergoes ligand displacement instead and gives (CIV).

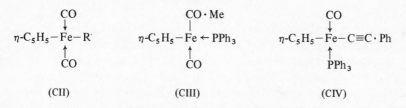

(CII)   (CIII)   (CIV)

With most alkylmetal carbonyls a decrease in the electron density on the metal enhances the rate at which the alkyl group migrates on to an adjacent carbon monoxide ligand. This is particularly true if the decrease is the result of an increase in the formal oxidation state of the metal, and thus the one-electron oxidation (e.g., by Ce^{4+}) of complexes of the types R . Mo(η-C$_5$H$_5$)(CO)$_3$ and R . Fe(η-C$_5$H$_5$)(CO)$_2$ is immediately followed by migration of the group R.

As expected for an intramolecular migration, essentially complete retention of configuration is observed in the formation of R . CO . Mn(CO)$_5$ by the action of carbon monoxide on R . Mn(CO)$_5$ in which the group R is bonded to the metal by an asymmetric carbon. The same stereochemical result is obtained when R . Fe(η-C$_5$H$_5$)(CO)$_2$ is converted into R . CO . Fe (η-C$_5$H$_5$)(CO)(PPh$_3$) by the action of triphenylphosphine.

In the examples that have been discussed so far, the carbon monoxide that becomes part of the acyl group is co-ordinated with the metal in the starting complex. This need not necessarily be so, for some metal–alkyl complexes afford acyl derivatives when treated with carbon monoxide because they initially give an unstable metal carbonyl which then rearranges.

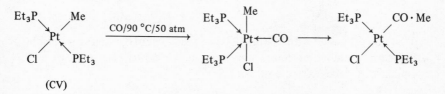

(CV)

This formation of a metal acyl from a metal alkyl and carbon monoxide is very important synthetically, for it is an essential step in many of the metal-catalysed syntheses of carbonyl compounds in which the carbonyl group is derived from carbon monoxide (see §5.8).

Similarly the carbon monoxide ligand that is formed when the alkyl group migrates from the acyl group to the metal need not remain co-ordinated, but may be evolved as free carbon monoxide. The overall reaction then constitutes decarbonylation of the —CO . R group.

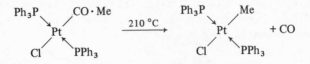

The ease with which metal alkyls can be carbonylated, and with which metal acyls can be decarbonylated, depends upon the metal, its oxidation state, and the ligands with which it is co-ordinated. Thus in contrast to the forcing conditions required to carbonylate the platinum complex (CV), the corresponding palladium complex can be carbonylated at room temperature and under atmospheric pressure because of the greater ease with which palladium(II) increases its co-ordination number to five. Some comparative rates of decarbonylation are given on p. 43. The processes of carbonylation and decarbonylation have been reviewed in detail by Wojcicki (1973) and Calderazzo (1977).

Compared with metal carbonyls, complexes of organic isonitriles ($\text{C} \equiv \text{N}^+ -\text{R}$) have received relatively little attention. A useful source of information on these complexes is the book by Malatesta and Bonati (1969), and the review by Treichel (1973). Simple isonitriles are weaker π-acceptors than is carbon monoxide but stronger σ-donors, and with metals of the d^8 configuration they appear to be among the most powerful unidentate ligands known. Thus methyl isocyanide displaces all the ligands in $IrCl(CO)(PPh_3)_2$ and $[Ir(CO)(pdma)_2]^+$ to give $[Ir(CN.Me)_4]^+$. Like carbon monoxide, co-ordinated isonitriles are attacked by nucleophiles particularly when there is only a low degree of back-bonding from the metal. Cationic iron complexes, for example, react with organolithium reagents to give iminacyl derivatives.

When co-ordinated with platinum(II) and palladium(II), isonitriles react with alcohols, and with primary and secondary amines to give stable carbene complexes, as in the example shown.

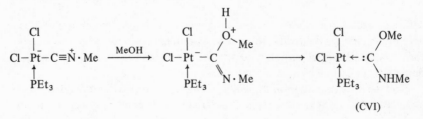

(CVI)

These reactions are of interest in that in the unco-ordinated state the carbene formed would be expected rapidly to rearrange, e.g. the carbene in the complex (CVI) would be expected to give the formamidate ester (CVII; $R = R^1 = Me$). As indicated by equation 5.8, esters of this type are actually formed when the appropriate isonitrile and alcohol are heated together in the presence of metal catalysts such as metallic copper, or the halides of copper(I), silver(I) and mercury(II). This suggests that in the metal-catalysed reactions, the

$$R \cdot \overset{+}{N}\equiv C^- + R^1 \cdot OH \xrightarrow{\ CuCl\ } R \cdot N = CH \cdot OR^1 \qquad [5.8]$$
$$\text{(CVII)}$$

$$R - \overset{+}{N}\equiv C^- + R^1R^2NH \xrightarrow{\ AgCl\ } R \cdot N = CH \cdot NR^1R^2 \qquad [5.9]$$

ester formation may proceed *via* an intermediate carbene complex formed by nucleophilic attack on a complex of the isonitrile. The metal-catalysed formation of formamidines [5.9] may be rationalised in a similar manner. This idea is supported by the structure of the products obtained when isonitriles react in the presence of copper(I) or silver(I) with ethylamines that have a X—H group (X = O, S, or NR) in the 2-position. This reaction gives unsaturated five-membered heterocycles, whose formation may be rationalised in terms of a complexed carbene which undergoes an intramolecular insertion into the X—H bond.

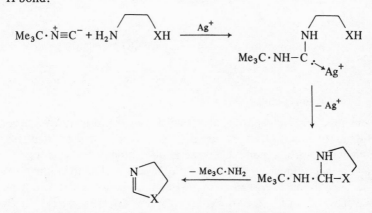

Several examples of the migration of an alkyl group from a transition metal to an adjacent co-ordinated isonitrile ligand are known. When the salt (CIX) formed from methyl isonitrile and the platinum complex (CVIII) is refluxed in benzene, the iminoacyl derivative (CX) is obtained. This particular reaction sequence results in the insertion of the isonitrile into a metal–carbon bond, and is directly analogous to the carbonylation of the complex (CV) (p. 271).

A similar insertion is involved in the use of isonitriles for preparing *o*-substituted aromatic aldehydes *via* aryl–copper(I) compounds in which

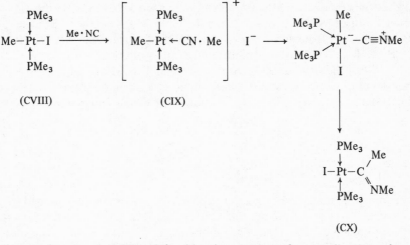

(CVIII) (CIX) (CX)

the metal–carbon bond is stabilised by the presence of a co-ordinating substituent in the *ortho*-position (cf. p. 66).

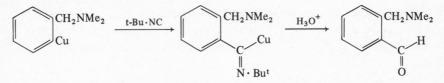

Insertions of isonitriles have been reviewed by Yamamoto and Yamazaki (1972).

5.8.   Carbonylation

There are a large number of syntheses of aldehydes, ketones and carboxylic acids and their derivatives that are catalysed or promoted by transition metals, and in which carbon monoxide is the source of the carbonyl group in the final organic product. In all of these syntheses, or 'carbonylations' as they are collectively termed, the carbon monoxide is in the form of metal carbonyls which are either used as starting materials or which are generated during the reaction.

Although most carbonylations are essentially 'one-pot' reactions, the majority involve three distinct, successive stages:

(1) the formation of a complex in which one of the ligands is σ-bonded to the metal through carbon,

(2) the formation of a metal acyl by the migration of this ligand on to adjacent co-ordinated carbon monoxide, and

(3) the cleavage of the metal acyl to give a metal-free organic carbonyl compound.

The second of these three stages has already been discussed in the preceding section, and for convenience the last stage will now be briefly considered first, using examples which in some cases also illustrate two minor methods, i.e. oxidative addition and nucleophilic attack on a co-ordinated alkene, for producing the σ-bonded ligand in the first stage. It must be pointed out, however, that most of the carbonylations that are used as examples – and, in fact, most carbonylations – have not been studied mechanistically, and the mechanisms that have been proposed are based largely on analogy with reactions of stable metal carbonyls, particularly those of cobalt and manganese. This is true, for example, of those carbonylations that are promoted or catalysed by palladium (see review by Tsuji, 1969, and Trost, 1977), for very few palladium carbonyls have been characterised and comparatively little is known about their chemistry.

(i) *Cleavage of metal acyls*

The stability of certain anionic metal complexes that contain a number of strong π-acceptor ligands such as carbon monoxide enables these species to act as leaving groups in substitutions and eliminations (see p. 39), and they can be displaced from an acyl residue by nucleophilic attack at the carbonyl carbon atom. Specific examples are provided by the reactions of manganese and cobalt acylcarbonyls with amines, alcohols and alkoxide anions.

$$\text{EtO}^- \curvearrowright \quad \underset{\underset{O}{\|}}{\text{Me}-\text{C}-\text{Mn(CO)}_5} \longrightarrow \text{Me}\cdot\text{CO}_2\text{Et} + [\text{Mn(CO)}_5]^-$$

The cleavage of cobalt acyls by this particular route occurs in the final stage of an important process for preparing carboxylic esters from alkyl halides (see p. 280). If the metal does not have strong π-acceptor ligands its ability to bear a negative charge is decreased, and nucleophilic attack on the acyl residue may still take place, but it is then accompanied by a two-electron reduction of the metal and loss of an anionic ligand.

$$\text{Nu:} \curvearrowright \quad \underset{\underset{O}{\|}}{\text{R}-\text{C}-\text{M}-\text{L}} \longrightarrow \text{R}\cdot\text{CO}\cdot\text{Nu} + \text{M:} + \text{L:}$$

This type of acyl cleavage is particularly associated with platinum and palladium acyls, and is illustrated by the reaction of the complex (CXI) with methanol (see p. 276).

An example of a carbonylation in which this reductive cleavage is involved is the conversion of alkenes into β-alkoxyesters by the action of an alcohol,

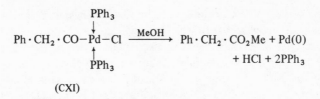

(CXI)

carbon monoxide and palladium(II) chloride, e.g. the conversion of *cis*-2-butene into the *threo*-methyl ester (CXIII). In the presence of carbon monoxide the unstable σ-bonded complex (CXII) initially produced by *trans*-attack of methanol on the co-ordinated alkene (cf. p. 203) is converted into a palladium acyl which is then reductively cleaved by methanol to give the observed ester.

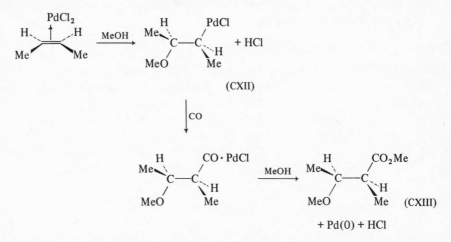

Closely related to the intermolecular reaction is the reductive elimination [5.10] in which one of the σ-bonded ligands can be regarded as acting as an internal nucleophile.

$$R-\underset{\underset{O}{\|}}{C}-M \longrightarrow R \cdot CO \cdot L + M: \qquad [5.10]$$

An example is the formation of the acyl chloride when the complex (CXIV; R = Me(CH$_2$)$_{14}$) is heated with carbon monoxide.

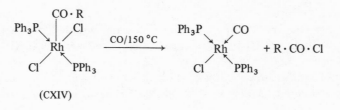

(CXIV)

This particular reaction is a good model for the elimination which is frequently proposed as the final step in carbonylations that result in the formation of carboxylic acid halides, for example the carbonylation of the nickel(II) complexes produced by the oxidative addition of aryl halides to certain *tert*-phosphine complexes of nickel(0).

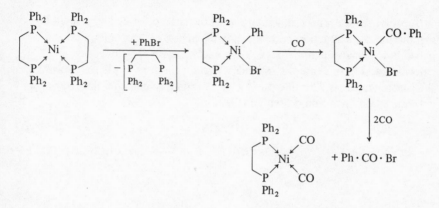

An analogous carbonylation in the aliphatic series is the formation of $\beta\gamma$-unsaturated acyl halides from π-allyl complexes of nickel(II) or palladium(II) halides.

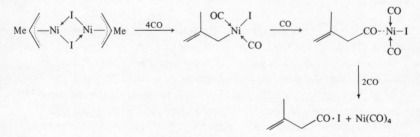

In the case of the palladium complexes this particular carbonylation can be made into a catalytic procedure for converting allylic halides into $\beta\gamma$-unsaturated acyl halides (equation 5.11), for in the presence of the allylic halide the palladium species produced in the last stage of the reaction is converted back into the starting π-allyl complex.

$$\text{CH}_2=\text{CH}-\text{CH}_2-\text{Cl} + \text{CO} \xrightarrow{\text{PdCl}_2} \text{CH}_2=\text{CH}-\text{CH}_2-\text{CO}\cdot\text{Cl} \qquad [5.11]$$

When this reaction is carried out using an alcohol as the solvent the expected carboxylic ester is produced, but it is not known whether this is formed directly by nucleophilic attack on the intermediate metal acyl by the solvent (see above), or indirectly *via* the acyl halide. The same ambiguity also applies to the related palladium(II)-catalysed conversion of aryl, heterocyclic and alkenyl halides into amides, as in the following example.

$$
\underset{S}{\text{(thienyl)}}Br + H_2NPh \xrightarrow{\text{PdBr}_2(\text{PPh}_3)_2/\text{CO}} \underset{S}{\text{(thienyl)}}CO \cdot NHPh
$$

Another carbonylation in which acid chlorides are formed by a reductive elimination is the reaction between simple alkenes, carbon monoxide and palladium(II) chloride in benzene. This reaction affords β-chloroacyl chlorides, and in this case the σ-bonded ligand produced in the first stage of the carbonylation is a β-chloroalkyl group formed by nucleophilic attack by chloride anion on the co-ordinated alkene.

$$
Me \cdot HC{=}CH_2 \xrightarrow[\text{}]{2CO} Me \cdot CHCl \cdot CH_2{-}\overset{CO}{\underset{CO}{Pd}}{-}Cl \xrightarrow{CO} Me \cdot CHCl \cdot CH_2 \cdot CO{-}\overset{CO}{\underset{CO}{Pd}}{-}Cl
$$

$$
\downarrow
$$

$$
Me \cdot CHCl \cdot CH_2 \cdot CO \cdot Cl + Pd(0) + 2CO
$$

Here again it is possible to obtain the corresponding ester if alcohol is the solvent.

In the carbonylations used above to illustrate the reductive elimination [5.10] the ligand (L), i.e. the halide anion, was present on the metal before the acyl group was formed. In some carbonylations the ligand L becomes attached to the metal, e.g. by a substitution or by an oxidative addition, after the acyl group has been formed. A specific example of this situation is provided by the process of hydroformylation (see p. 283) in which the oxidative addition of dihydrogen to a cobalt acyl is followed by the reductive elimination of an aldehyde.

$$
R \cdot CO \cdot Co(CO)_3 \xrightarrow{H_2} R \cdot CO \cdot \overset{H}{\underset{H}{Co}}(CO)_3 \longrightarrow R \cdot CHO + CoH(CO)_3
$$

Another general method for cleaving metal acyls (M . CO . R) is by treatment with a halogen (X_2). This affords the corresponding metal halide (MX) and the acyl halide (R . CO . X), and the latter may be isolated either as such or as a carboxylic acid, ester, or amide after further reaction with water, an alcohol, or an amine respectively. A good illustration of this particular method of cleavage is provided by a general stereospecific synthesis of β-lactams – a carbonylation in which the σ-bonded metal alkyl produced in the first stage is formed by nucleophilic attack on a co-ordinated alkene (again, by *trans*-attack).

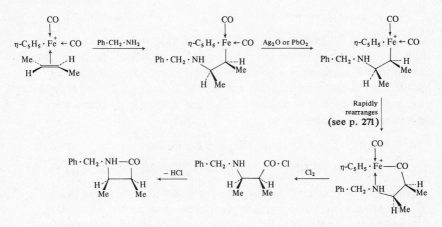

Having briefly discussed some general processes by which an acyl residue may be cleaved from the metal in the final stage of a carbonylation, some specific carbonylations will now be considered in greater detail. In all of the carbonylations the three stages mentioned on p. 274 can be identified, and for convenience the carbonylations are divided into two groups that differ from each other in the way in which a σ-bonded ligand is formed in the first stage.

(ii)   *Formation of σ-bonded organic ligands by nucleophilic attack with anionic metal complexes*

One way by which σ-bonded organic ligands can be formed in the first stage of a carbonylation is by nucleophilic attack of an anionic metal carbonyl on a carbon atom that bears a suitable leaving group.

$$X-R \quad M \leftarrow CO \longrightarrow X^- + R-M \leftarrow CO$$

The most clear-cut examples of this type of nucleophilic substitution are provided by reactions of the tetracarbonylcobalt(-I) anion $[Co(CO)_4]^-$, which in methanol has a nucleophilicity (for carbon) of the same order as that of the methoxide anion. The cobalt anion is conveniently prepared by reduction of octacarbonyldicobalt with lithium or sodium amalgam, and reacts with alkyl halides to give the expected tetracarbonylcobalt alkyl.

$$Co_2(CO)_8 \xrightarrow{2Na} Na^+[Co(CO)_4]^- \xrightarrow{RX} R \cdot Co(CO)_4 + NaX$$

These alkyls, which are yellow to red in colour, are unstable and in the presence of organic ligands such as triphenylphosphine and carbon monoxide

readily arrange to give cobalt acyls, probably through intermediate acyltri-
carbonylcobalt complexes (cf. equations 5.6 and 5.7, p. 270).

$$R \cdot Co(CO)_4 \rightleftharpoons R \cdot CO \cdot Co(CO)_3 \xrightarrow{CO} R \cdot CO \cdot Co(CO)_4$$

Therefore if the reaction between an alkyl halide and the tetracarbonylcobalt anion is carried out in the presence of carbon monoxide an acyltetra-
carbonylcobalt complex is formed. Now as indicated on p. 275, the acyl-
tetracarbonylcobalt complexes are susceptible to nucleophilic attack, and they
react with primary and secondary alcohols and amines to give carboxylic
esters and amides respectively.

$$R \cdot CO \cdot Co(CO)_4 + R'OH \rightarrow R \cdot CO_2R' + CoH(CO)_4$$

$$R \cdot CO \cdot Co(CO)_4 + 2R'NH_2 \rightarrow R \cdot CONHR' + [R' \cdot NH_3]^+ [Co(CO)_4]^- \quad [5.12]$$

Consequently if an alkyl halide is treated with the tetracarbonylcobalt
anion and carbon monoxide with an alcohol as the solvent, the acyltetra-
carbonylcobalt complex that is formed is solvolysed to give a carboxylic
ester. The overall reaction is therefore:

$$RX + CO + R'OH + [Co(CO)_4]^- \rightarrow R \cdot CO_2R' + CoH(CO)_4 + X^-$$

In the presence of a suitable base, e.g. the alkoxide anion $(R'O^-)$ or a
tertiary amine, the hydridotetracarbonylcobalt formed is converted back
into the tetracarbonylcobalt anion, thus allowing the formation of the ester
from the alkyl halide to be carried out using a catalytic rather than a stoichiometric amount of the cobalt anion. The anion is also reformed when tetracarbonylcobalt acyls are treated with more than one equivalent of a primary
or secondary amine (see equation 5.12) and hence under these conditions the
formation of amides is automatically catalytic with respect to the hydridotetracarbonylcobalt anion.

Tetracarbonylcobalt acyls react very rapidly with iodine to give high yields
of the acyl iodide.

$$2R \cdot CO \cdot Co(CO)_4 + 3I_2 \rightarrow 2R \cdot CO \cdot I + 2CoI_2 + 4CO$$

If this reaction is carried out in an alcohol the expected carboxylic ester is
formed from the acyl iodide. This provides an alternative method for converting
tetracarbonylcobalt acyls into carboxylic esters, and is particularly useful when
it is desirable or necessary initially to prepare the cobalt acyl in a non-hydroxy-
lic solvent, as in the following reaction.

Oxiranes react with the tetracarbonylcobalt anion and also with the parent
hydride, thus providing a preparation of β-hydroxyesters.

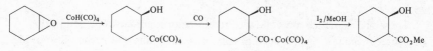

The faster rates of reaction that are associated with the parent hydride (which can function as a strong acid, see p. 40), and the structures of the alcohols obtained from unsymmetrical oxiranes, suggest that the reactions with this reagent involve nucleophilic attack on the protonated oxirane.

A further synthetic application of the nucleophilic tetracarbonylcobalt anion is the so-called 'acyldiene synthesis' by which an alkyl chain can be extended by a dienone residue.

$$RX + CO + H\overset{|}{C} = \overset{|}{C} - \overset{|}{C} = \overset{|}{C} - \rightarrow R \cdot CO \cdot \overset{|}{C} = \overset{|}{C} - \overset{|}{C} = \overset{|}{C} - + HX$$

In this process (see Heck, 1968, for discussion) an alkyl halide or tosylate is subjected to nucleophilic attack by the cobalt anion in the presence of a conjugated diene to give an acyldiene complex which rearranges to an acyl π-allyl complex (CXV) (drawn for convenience in the σ-allyl form). This rearrangement appears to involve the migration on to the co-ordinated diene of the acyl group, i.e. a group that in organometallic chemistry usually has a very low migratory aptitude. When treated with a base, preferably a hindered tertiary amine, the acyl π-allyl complex forms a dienone by undergoing a β-elimination (see arrows, (CXV)) with the metal carbonyl residue as the leaving group.

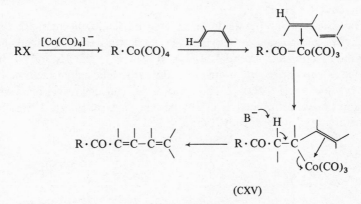

(CXV)

The isoelectronic (d^{10}) dianion, $[Fe(CO)_4]^{2-}$, can be used in a manner analogous to the use of the cobalt monoanion, and is prepared *in situ* by reduction of pentacarbonyliron(0) with sodium amalgam. The dianion reacts with primary and secondary alkyl halides and tosylates by an S_N2 mechanism (inversion of configuration at carbon) to give anionic iron alkyls, which with carbon monoxide rearrange by alkyl migration to give the corresponding iron acyls. These acyls can also be prepared by direct acylation of the tetracarbonyl-iron dianion (see p. 282).

Like the corresponding cobalt acyls, the iron acyls are cleaved by iodine to give acyl iodides, and if this is carried out in the presence of water or alcohols,

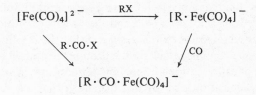

carboxylic acids or esters respectively are formed. An additional synthetic advantage of the iron acyls, however, is that because they are negatively charged they can function as nucleophiles, and with reactive alkyl halides they afford unstable iron(II) complexes which decompose *in situ* to give ketones by reductive elimination.

$$[R \cdot CO \cdot Fe(CO)_4]^- \xrightarrow{\;R'X\;} R \cdot CO - \overset{\overset{\displaystyle R'}{|}}{Fe}(CO)_4 \longrightarrow R \cdot CO \cdot R' + Fe(CO)_4$$

(CXVI)

Protonation of the anionic iron acyls, for example with acetic acid, affords unstable hydrides (CXVI; R' = H) which decompose similarly to give aldehydes (R . CHO).

An advantage of using the tetracarbonyliron dianion in the synthesis of aldehydes, ketones and carboxylic acids and derivatives as described above is that the functional groups CHO, CO_2R, R . CO . R', CN, NH_2 and OH in the residue (R) do not require protection. The carboxylic anhydride function is reactive towards $[Fe(CO)_4]^{2-}$, however, and cyclic anhydrides react with the dianion to produce ultimately aldehydic acids in excellent yield, presumably *via* an intermediate of type (CXVI; R' = H).

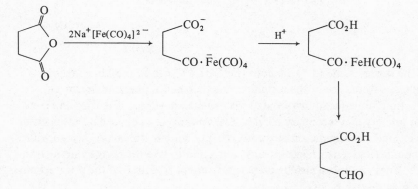

Synthetic applications of the $[Fe(CO)_4]^{2-}$ anion have been discussed by Collman (1975).

(iii)   *Formation of σ-bonded organic ligands from π-bonded alkenes and alkynes*

Another important route by which a σ-bonded organic ligand can be formed in the first stage of a carbonylation is by a hydride ligand migrating from the metal on to a co-ordinated alkene.

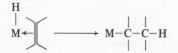

This process, which has been discussed in general terms in §5.4 and in connection with catalytic hydrogenations in §5.5, forms a basic part of the Oxo (or Roelen) reaction in which an alkene reacts with carbon monoxide and hydrogen in the presence of a metal catalyst to give an aldehyde.

$$\begin{array}{c}\diagdown\\ /\end{array} C = C \begin{array}{c}/\\ \diagdown\end{array} + CO + H_2 \longrightarrow H - \underset{|}{\overset{|}{C}} - \underset{|}{\overset{|}{C}} - CHO$$

As this carbonylation affords a product in which H—CHO has been added across the double bond of the starting alkene it is often referred to as 'hydroformylation'. The reaction is of great industrial importance for producing aldehydes in the C_3–C_{18} range, which can be oxidised to the corresponding carboxylic acids or – more commonly – reduced to the primary alcohol. This reduction is usually carried out as a separate step, but some industrial plants operate under conditions that ensure that the aldehyde is reduced *in situ*. Of the many metals and inorganic compounds that can function as catalysts in hydroformylations, cobalt carbonyls or cobalt compounds that can form carbonyls under the conditions of the reaction are most commonly used. With stoichiometric amounts of cobalt carbonyls the reaction proceeds at room temperature and atmospheric pressure, but on the industrial scale where only catalytic amounts are used temperatures in excess of 100 °C and pressures of up to 300 atmospheres are necessary. In the process developed by Shell Ltd, the catalysts are mixed phosphine cobalt carbonyls such as $[\text{n-Bu}_3\text{P}.\text{Co(CO)}_3]_2$, and these have the advantage that they work efficiently at low pressures (5–35 atmospheres, 200 °C) and give alcohols directly. Rhodium compounds have recently been found to be very effective as catalysts, and are often up to 10^4 times more efficient than cobalt carbonyls.

On account of its industrial importance hydroformylation has received substantial attention from the mechanistic viewpoint. It is now generally accepted that in the cobalt-catalysed systems the active catalyst is hydridotetracarbonylcobalt, which is formed *in situ* by reduction of the cobalt carbonyls produced by the action of carbon monoxide on the cobalt or cobalt compounds initially added to the reaction mixture.

$$2Co + 8CO \longrightarrow Co_2(CO)_8 \xrightarrow{\text{H}_2} 2CoH(CO)_4$$

A dissociative mechanism allows one of the carbon monoxide ligands in the hydridotetracarbonylcobalt to be replaced by the alkene, and the migration of the hydride ligand on to the co-ordinated alkene generates a carbon-bonded alkyl group.

$$CoH(CO)_4 \underset{-CO}{\rightleftarrows} CoH(CO)_3 \xrightarrow{H_2C=CH_2} H-Co(CO)_3 \longrightarrow Et \cdot Co(CO)_3$$

(with $H_2C=CH_2$ shown above $H-Co(CO)_3$)

Subsequent migration of this group on to co-ordinated carbon monoxide affords a tricarbonylcobalt acyl, which is co-ordinatively unsaturated and can add either a further molecule of carbon monoxide or dihydrogen.

In the latter case the addition is followed by a reductive elimination of an aldehyde to leave hydridotricarbonylcobalt which can then initiate a fresh cycle of reactions.

$$Et \cdot Co(CO)_3 \xrightarrow{CO} Et \cdot Co(CO)_4 \longrightarrow Et \cdot CO \cdot Co(CO)_3$$

$$\downarrow H_2 \qquad \qquad \qquad \qquad$$

$$Et \cdot CHO + CoH(CO)_3 \longleftarrow Et \cdot CO \cdot Co(CO)_3 \text{ (with H above and H below)}$$

On the basis of studies made by Wilkinson and coworkers at Imperial College, it seems certain that rhodium-catalysed hydroformylations also involve a hydridocarbonyl complex as the active catalyst. Wilkinson found that the stable tris(triphenylphosphine) complex, $RhH(CO)(PPh_3)_3$, very efficiently catalyses the hydroformylation of alkenes even at room temperature and atmospheric pressure, and that the reaction involves the initial formation of a $Rh(alkene)(H)(CO)_2(PPh_3)_2$ complex which undergoes the same sequence of reactions as that postulated for the cobalt systems, i.e. formation of a metal-bonded alkyl group which migrates on to an adjacent carbon monoxide ligand, followed by oxidative addition of dihydrogen to the metal and reductive elimination of the aldehyde. In phosphine-free rhodium systems the hydridocarbonyl complex, $RhH(CO)_4$, presumably acts as the catalyst in an identical manner.

One of the early steps in rhodium- and cobalt-promoted hydroformylations is the formation of a metal alkyl by migration of a hydride ligand on to the co-ordinated alkene. As described on p. 242, the stereochemistry of this migration is consistent with *syn*-addition of M—H across the carbon–carbon double bond. As the configuration of the carbon atom attached to the metal in the metal alkyl is retained in all the subsequent steps of the hydroformylation the stereochemistry of the overall conversion of alkene into saturated aldehyde is also consistent with *syn*-addition (of H—CHO) to the carbon–carbon double bond. Hydroformylation of the *E*-alkene (CXVII), for example, affords the racemic *threo*-alcohol (CXVIII) with 95 per cent stereoselectivity.

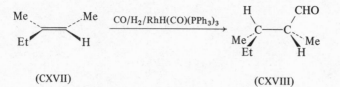

(CXVII)   (CXVIII)

A further consequence of the migration of a hydride ligand from the metal on to the double bond of the co-ordinated alkene is that the hydroformylation of unsymmetrical alkenes can afford two isomeric aldehydes.

$$\begin{array}{c} MeCH{=}CH_2 \\ \downarrow \\ H{-}Co(CO)_3 \end{array} \Bigg\langle \begin{array}{l} Me_2CH\cdot Co(CO)_3 \longrightarrow Me_2CH\cdot CHO \\ \\ n\text{-}Pr\cdot Co(CO)_3 \longrightarrow n\text{-}Pr\cdot CHO \end{array}$$

In practice more than two isomers are usually obtained, for the initial tricarbonylcobalt alkyl can isomerise by undergoing reversible elimination/addition of hydridotricarbonylcobalt (cf. p. 263). Thus the hydroformylation of 3-^{2}H-3-methyl-1-hexene (CXIX), p. 287, affords not only the expected aldehydes (CXX), 91.5 per cent, and (CXXI), 5.0 per cent, but also the isomers (CXXII), 2.3 per cent, and (CXXIII), 0.7 per cent, the last two compounds being formed as shown in figure 5.5. Under the industrial conditions normally used the hydroformylation of linear, terminal alkenes affords mainly linear aldehydes, the percentage amounts of which are increased by the use of high carbon monoxide pressures, relatively low temperatures, and phosphine-containing catalysts. This predominance of linear aldehydes reflects the greater stability of metal alkyls in which the metal is attached to a primary rather than a secondary or tertiary carbon, i.e. alkyls in which any steric hindrance between the alkyl chain and other ligands on the metal is minimal (see p. 243). This difference in stability between isomeric metal alkyls is convincingly shown by the stoichiometric reaction under laboratory conditions between hydridotetracarbonylcobalt and $\alpha\beta$-unsaturated esters. In the case of ethyl acrylate the initial product is the kinetically controlled one, i.e. the 2-substituted ester (CXXIV), but in solution this readily isomerises to the more stable 3-substituted isomer (CXXV). It is not surprising that by varying the reaction conditions, the hydroformylation of ethyl acrylate can be made to yield predominantly either the 2- or 3-formyl derivative of ethyl propionate.

Although alkynes can be hydroformylated in the same manner as alkenes, it is evident from the low yields of products reported in the literature that this is not a convenient route to $\alpha\beta$-unsaturated aldehydes.

The most comprehensive discussion on hydroformylation is that by Falbe (1970), who has also discussed at length the related carbonylations described below.

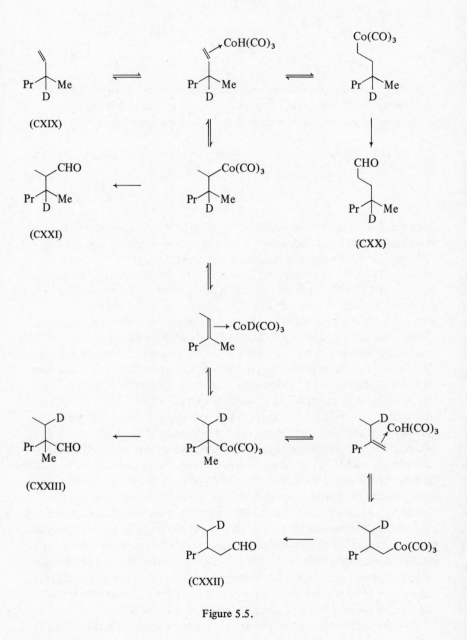

Figure 5.5.

$$\text{H}_2\text{C}=\text{CH}\cdot\text{CO}_2\text{Et} + \text{CoH(CO)}_3 \overset{\nearrow}{\underset{\searrow}{}}$$

$$\underset{(\text{CXXIV})}{\overset{\overset{\displaystyle\text{CO}_2\text{Et}}{|}}{\text{Me}\cdot\text{CH}\cdot\text{Co(CO)}_3}} \longrightarrow \overset{\overset{\displaystyle\text{CO}_2\text{Et}}{|}}{\text{Me}\cdot\text{CH}\cdot\text{CHO}}$$

$$\underset{(\text{CXXV})}{\overset{\overset{\displaystyle\text{CO}_2\text{Et}}{|}}{\text{CH}_2\cdot\text{CH}_2\cdot\text{Co(CO)}_3}} \longrightarrow \overset{\overset{\displaystyle\text{CO}_2\text{Et}}{|}}{\text{CH}_2\cdot\text{CH}_2\cdot\text{CHO}}$$

In the presence of metal carbonyls, alkynes react with a mixture of carbon monoxide and water to give $\alpha\beta$-unsaturated acids.

$$\text{HC}\equiv\text{CH} + \text{CO} + \text{H}_2\text{O} \xrightarrow{\text{Ni(CO)}_4} \text{H}_2\text{C}=\text{CH}\cdot\text{CO}_2\text{H}$$

The acids obtained from mono- and di-substituted alkynes by this process of 'hydrocarboxylation' may be regarded as having been formed by the *syn*-addition of H—CO$_2$H across the triple bond, with the hydrogen usually having become preferentially attached to the carbon atom that can be protonated more readily, i.e. the addition occurs in the Markownikov sense.

$$\text{AcO}\cdot\text{CH}_2\cdot\text{C}\equiv\text{C}\cdot\text{CH}_2\cdot\text{OAc} \longrightarrow \underset{\text{H} \quad\quad \text{CO}_2\text{H}}{\overset{\text{AcO}\cdot\text{CH}_2 \quad \text{CH}_2\cdot\text{OAc}}{\text{C}=\text{C}}}$$

$$\text{n-Bu}\cdot\text{C}\equiv\text{CH} \longrightarrow \text{n-Bu}\cdot\text{CH}\overset{cis}{=\!=\!=}\text{CH}\cdot\text{CO}_2\text{H} + \underset{\overset{|}{\text{CO}_2\text{H}}}{\text{n-Bu}\cdot\text{C}=\text{CH}_2}$$

$$\quad\quad\quad\quad\quad\quad\quad\quad \text{3.5 per cent} \quad\quad\quad\quad \text{35 per cent}$$

Hydrocarboxylation was one of the many carbonylations discovered and examined by Reppe and coworkers during the Second World War in the laboratories of Badische Anilin und Soda-Fabrik.

Of the various carbonyls that were found to be effective, tetracarbonyl-nickel(0) is the preferred one for use under laboratory conditions, and is employed in stoichiometric amounts at atmospheric pressure and temperatures below 80 °C. Under industrial conditions (120–250 °C; 20–30 atmospheres) only a catalytic amount of the carbonyl is required, and this is usually generated *in situ* from a nickel salt. The use of phosphine ligands and compounds of rhodium and palladium allows the catalytic reaction to proceed under less vigorous conditions than when nickel compounds are used.

In the stoichiometric method with tetracarbonylnickel(0), hydrochloric, phosphoric or, preferably, acetic acid must be included in the reaction mixture. It seems likely that the function of this acid (HX) is to convert an initially-formed alkyne π-complex into a σ-bonded alkenylnickel species. As mentioned on p. 228, one route by which this conversion can occur is by migration of a hydride ligand on to the co-ordinated alkyne after an initial oxidative addition.

$$\begin{array}{c} CH \\ \parallel \\ CH \end{array} \longrightarrow Ni(CO)_3 \longrightarrow \begin{array}{c} CH \\ \parallel \\ CH \end{array} \longrightarrow \overset{\overset{X}{|}}{\underset{\underset{H}{|}}{Ni(CO)_3}} \longrightarrow H_2C=CH-\overset{\overset{X}{|}}{Ni(CO)_3}$$

$$H_2C=CH \cdot CO \cdot NiX(CO)_2$$

(CXXVI)

Whether this is the actual route or not in the case of nickel complexes is still unknown. Very little is also known about the way in which the final unsaturated acid is formed from the intermediate nickel acyl (represented as the dicarbonyl (CXXVI)). It has been suggested that the metal acyl undergoes nucleophilic attack at the acyl group by a water molecule. This seems doubtful, however, for if the carbonylation is carried out using an alcohol as the solvent the major product is still the unsaturated acid rather than the corresponding ester. Small amounts of the ester are formed, but these appear to arise by esterification of the acid under the acidic conditions rather than by solvolysis of the metal acyl. Indeed from the mechanistic point of view it is very significant that 2-phenylacrylic acid can be obtained in as much as 20 per cent yield when phenylacetylene is treated with tetracarbonylnickel(0) and glacial acetic acid in dry anisole, i.e. under anhydrous conditions.

Hydrocarboxylation of alkenes to give saturated carboxylic acids can also be accomplished by use of metal carbonyls, but the conditions necessary are far more vigorous than those used in the hydrocarboxylation of alkynes. Even with stoichiometric amounts of tetracarbonylnickel(0), pressures of up to 50 atmospheres (at 150 °C) are necessary, while use of catalytic amounts of the carbonyl requires pressures and temperatures of up to 200 atmospheres and 250 °C. Photochemical activation, however, allows the reaction to proceed under normal laboratory conditions.

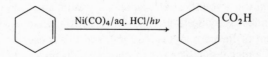

The difference in the reactivity of alkenes and alkynes under purely thermal conditions has proved to be of use in organic synthesis, e.g. in the preparation of the acid (CXXVII).

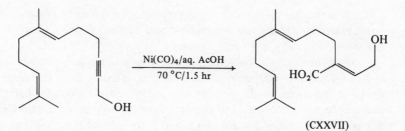

(CXXVII)

In the catalytic procedure for hydrocarboxylating alkenes and alkynes the water may be replaced by alcohols, amines and thiols; the products are then carboxylic esters, amides and thiol esters respectively.

$$Ph \cdot C{\equiv}CH + CO + EtSH \longrightarrow Ph-\underset{\underset{CH_2}{\|}}{\overset{\overset{CO \cdot SEt}{|}}{C}}$$

$$H \cdot C{\equiv}C \cdot H + CO + PhNH_2 \longrightarrow H_2C{=}CH \cdot CO \cdot NHPh$$

These reactions can be adapted for preparing 5- and 6-membered heterocycles by using alkenes or alkynes that contain an OH, NH_2, or SH group in the appropriate position. Both the stoichiometric and the catalytic methods can be applied.

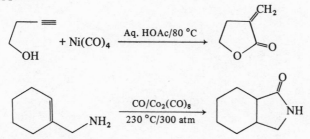

It should be noted that a carbonylation procedure can also be used for preparing heterocycles from starting materials of the type (CXXIX; X = CH, C . alkyl, C . aryl, or N; Y = Ar, OH, NHAr, or $NH . CO . NH_2$). Thus, 2-phenylphthalimidine (CXXX), 84 per cent, and 2-phenyl-3-indazolinone (CXXVIII), 55 per cent, can be obtained by carbonylating benzylideneaniline and azobenzene (CXXIX; Y = Ph; X = CH and N respectively) at 170–200 °C and 150–200 atmospheres with $Co_2(CO)_8$ as the catalyst.

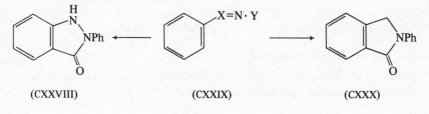

(CXXVIII)   (CXXIX)   (CXXX)

These and other preparations of heterocycles by the carbonylation of various starting materials have been discussed by Heck (1968), Thompson and Whyman (1971), and Bird (1973).

In the hydroformylation – and possibly also in the hydroxycarboxylation – of alkenes and alkynes the migration of a hydride ligand from the metal on to the π-bonded alkene or alkyne converts the latter into a σ-bonded ligand, which in turn migrates on to an adjacent co-ordinated carbon monoxide to give an acyl derivative. An example in which the σ-bonded organic ligand appears to be formed from a π-bonded alkyne by the migration of another ligand already bonded through *carbon* is provided by the reaction which takes place in an aqueous medium between allylic chlorides, acetylene and tetracarbonylnickel(0), and which affords *cis*-2,5-dienoic acids. It has been suggested that an initial oxidative addition of the allylic chloride gives a π-allyl complex which is converted into a σ-allyl complex when the metal co-ordinates with the acetylene. Migration of the allyl residue on to the co-ordinated acetylene is then followed by migration of the resultant *cis*-alkenyl residue on to carbon monoxide.

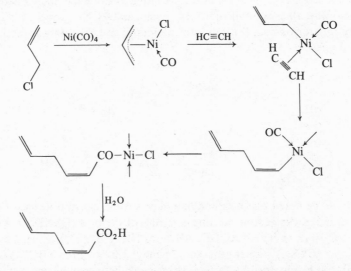

In a convenient modification of this reaction, a nickel catalyst prepared by *in situ* reduction of nickel(II) chloride with a manganese–iron alloy in the presence of thiourea can be used to effect the preparation of methyl *cis*-2,5-dienoates from allylic chlorides, acetylene, carbon monoxide and methanol in high yield at room temperature and atmospheric pressure.

(iv)   *Other carbonylation routes*

In the preceding discussion a number of carbonylations have been rationalised as three-stage processes, the first stage involving the formation of a complex with an organic ligand σ-bonded through carbon. There are some carbonylations, however, which although occurring in three distinct stages, differ from those described so far with respect to the type of processes that occurs in each of the three stages. In these carbonylations the first stage involves nucleophilic attack (either intra- or inter-molecular) on co-ordinated carbon monoxide, and the resultant metal complex reacts in the second stage to cause the attachment of a σ-bonded organic ligand, e.g. by oxidative addition or by functioning as a nucleophile. In stage three a reductive elimination affords the final metal-free organic carbonyl compound. The three stages can be summarised by the following general equation.

$$\text{M} \leftarrow \text{CO} \longrightarrow \bar{\text{M}}-\text{CO}-\text{Nu} \longrightarrow \overset{\overset{\displaystyle R}{\displaystyle |}}{\text{M}-\text{CO}-\text{Nu}} \longrightarrow \text{M:} + \text{R} \cdot \text{CO} \cdot \text{Nu}$$

Although the number of carbonylations that proceed by this route is small at the present time, it is clear that many more will be discovered as the chemistry of metal carbonyls is extended. Two factors that contribute to the synthetic potential of this route for carbonylation are the wide range of nucleophiles that can react with co-ordinated carbon monoxide, and the high reactivity of the resultant complexes which are usually anionic in nature. In the three examples briefly mentioned below, the nucleophiles involved are an alkoxide anion, hydroxyl anion, and a carbanion, and all three carbonylations are synthetically useful to the organic chemist.

The first example is the conversion of alkenyl halides into the corresponding carboxylic esters by the action of tetracarbonylnickel(0) and a metal alkoxide in the appropriate alcohol.

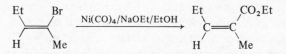

In this reaction the anionic complex $[\text{RO}.\text{CO}.\text{Ni(CO)}_3]^-$ is initially formed by nucleophilic attack on the nickel carbonyl by the alkoxide anion, and this reacts with the alkenyl halide ($\text{R}'\text{X}$) to give an unstable nickel(II) complex which affords the observed ester by reductive elimination of the alkenyl and $-\text{CO}.\text{OR}$ residues.

$$[\text{RO}\cdot\text{CO}\cdot\text{Ni(CO}_3)]^- \xrightarrow{\text{R}'\text{X}} \overset{\overset{\displaystyle R'}{\displaystyle |}}{\text{RO}\cdot\text{CO}\cdot\text{Ni(CO)}_3} \longrightarrow \text{RO}\cdot\text{CO}\cdot\text{R}'$$

One attraction of this particular carbonylation is that, as indicated in the example, the replacement of the halogen by the carboxylic ester group occurs stereospecifically.

Almost identical to the above reaction is the carbonylation of aryl halides to give aryl carboxylic acids.

$$ArX \xrightarrow{\text{Ni(CO)}_4/\text{CO}/\text{HO}^-} Ar \cdot CO_2H$$

This takes place at atmospheric pressure and at 100 °C when the halide is treated in an aprotic dipolar solvent, e.g. dimethylformamide, with tetra-carbonylnickel(0) and a source of hydroxide anions – preferably calcium hydroxide.

The preparation of 1,4-diketones from $\alpha\beta$-unsaturated ketones and the lithium salts produced from tetracarbonylnickel(0) and organolithium compounds (see p. 266) is another carbonylation that falls into the same class of reaction. In this case the organic ligand becomes σ-bonded to the metal in the second stage when the initial complex acts as the nucleophile in a 1,4-addition.

$$\text{n-Bu} \cdot \text{Li} + \text{Ni(CO)}_4 \longrightarrow [\text{n-Bu} \cdot \text{CO} \cdot \text{Ni(CO}_3)]^- \text{Li}^+$$

$$\Big\downarrow \text{Me}_2\text{C=CH}\cdot\text{CO}\cdot\text{Me}$$

$$\text{n-Bu} \cdot \text{CO} \cdot \text{Ni(CO)}_2$$
$$|$$
$$\text{CMe}_2 \cdot \text{CH=CMe}$$
$$|$$
$$\text{OLi}$$

$$\text{n-Bu} \cdot \text{CO} \cdot \text{CMe}_2 \cdot \text{CH}_2 \cdot \text{CO} \cdot \text{Me} \xleftarrow[\text{(2) H}_3\text{O}^+]{\text{(1) red. elim.}}$$

5.9.   Decarbonylation

In view of the reversibility of most of the reactions involved in metal-promoted carbonylations, it is not surprising that there are examples of decarbonylation, i.e. the removal of carbon monoxide from a carbonyl compound, that take place under the influence of transition-metal compounds.

It has been known for some time that in the presence of metallic palladium an aldehyde group can be selectively decarbonylated.

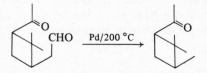

Recently it has been found that these decarbonylations can be accomplished under homogenous conditions in refluxing benzene, dichloromethane, or – in the case of sterically hindered aldehydes – benzonitrile, by use of the rhodium(I) complex, $RhCl(PPh_3)_3$.

$$R \cdot CHO + RhCl(PPh_3)_3 \longrightarrow RH + RhCl(CO)(PPh_3)_2 + PPh_3$$

(CXXXI)

If necessary the starting complex can be regenerated from the carbonyl derivative (CXXXI) by heating the latter with refluxing benzyl chloride and then treating the resultant complex, $RhCl_2(PPh_3)(Ph \cdot CH_2)$, with triphenyl-phosphine in ethanol.

The decarbonylation probably involves the oxidative addition of the aldehyde followed by alkyl migration from the acyl group to the metal, and then a reductive elimination of the hydrocarbon (RH).

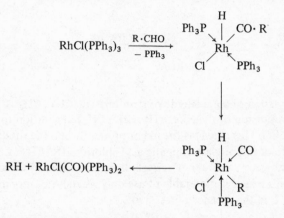

In accordance with this mechanism the decarbonylation of a deuteriated aldehyde ($R \cdot CDO$) affords the monodeuteriated hydrocarbon ($R \cdot D$), and the decarbonylation of optically active and $\alpha\beta$-unsaturated aldehydes, e.g. (CXXXII) and (CXXXIII), proceeds with a very high degree (> 90 per cent) of stereospecificity.

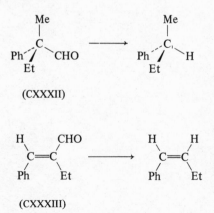

The selective decarbonylation of an aldehyde group by means of the rhodium(I) complex has been used in a number of synthetically useful transformations, e.g. in the removal of an aromatic methyl group after prior

oxidation with ammonium cerium(IV) nitrate, and in the stereospecific
formation of trisubstituted alkenes from naturally-occurring triterpenes.

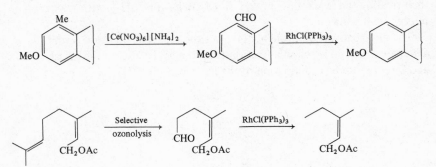

The decarbonylation of an aldehyde of structure $(R.CH_2.CH_2.CHO)$
affords a small proportion of the alkene $(R.CH = CH_2)$ in addition to the
hydrocarbon $(R.Et)$. This alkene is the major product when the rhodium
complex is used to decarbonylate aliphatic acyl chlorides $(R.CH_2.CH_2.COCl)$.
Aroyl chlorides, however, are very effectively decarbonylated to the aryl
chloride, but in this case it is preferable to use only a catalytic amount of the
complex at 260 °C in the absence of a solvent.

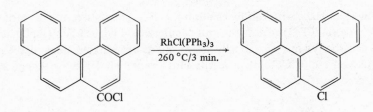

References

Abel, E. W. and Stone, F. G. A. (1969). *Quart. Rev.* **23**, 325.
Abel, E. W. and Stone, F. G. A. (1970). *Quart. Rev.* **24**, 498.
Ahrland, S., Chatt, J. and Davies, N. R. (1958). *Quart. Rev.* **12**, 265.
American Chemical Society (1974). *Advances in Chemistry Series*, no. 130.
Bacon, R. G. R. and Hill, H. A. O. (1965). *Quart. Rev.* **19**, 95.
Baker, R. (1973). *Chem. Rev.* **73**, 487.
Basolo, F., Hoffman, B. M. and Ibers, J. A. (1975). *Accounts Chem. Res.* **8**, 384.
Basolo, F. and Pearson, R. G. (1967). *Mechanisms of Inorganic Reactions* (New York: John Wiley and Sons).
Benkeser, R. A., Hooz, J., Liston, T. V. and Trevillyan, A. E. (1963). *J. Amer. Chem. Soc.* **85**, 3984.
Birch, A. J. and Jenkins, I. D. (1976). *Transition Metal Organometallics in Organic Synthesis*, vol. 1, p. 2. Ed. H. Alper (New York: Academic Press).
Birch, A. J. and Williamson, D. H. (1976). *Organic Reactions*, **24**, 1.
Bird, C. W. (1967). *Transition Metal Intermediates in Organic Synthesis* (London: Logos Press).
Bird, C. W. (1973). *J. Organomet. Chem.* **47**, 281.
Bjerrum, J., Schwarzenbach, G. and Sillen, L. G. (1957). *Stability Constants. Part I: Organic Ligands.* The Chemical Society, Special Publication no. 6.
Bjerrum, J., Schwarzenbach, G. and Sillen, L. G. (1958). *Stability Constants. Part II: Inorganic Ligands.* The Chemical Society, Special Publication no. 7.
Bloodworth, A. J. (1977). *The Chemistry of Mercury*, p. 139. Ed. C. A. McAuliffe (London: Macmillan).
Boyd, D. R. and McKervey, M. A. (1968). *Quart. Rev.* **22**, 95.
Bradley, D. C. (1960). *Prog. Inorg. Chem.* **2**, 303.
Brown, H. C. and Bartholomay, H. (1944). *J. Amer. Chem. Soc.* **66**, 435.
Bruce, M. I. (1977). *Angew. Chem. internat. Edit.* **16**, 73.
Bruce, M. I. and Stone, F. G. A. (1968). *Angew. Chem. internat. Edit.* **7**, 747.
Busch, D. H. (1967). *Helv. chim. Acta.* Alfred Werner Commemoration Volume, 174.
Cais, M. (1966). *Organomet. Chem. Rev.* **1**, 435.
Calderazzo, F. (1977). *Angew. Chem. internat. Edit.* **16**, 299.
Calderon, N. C., Ofstead, E. A. and Judy, W. A. J. (1976). *Angew. Chem. internat. Edit.* **15**, 401.
Candlin, J. P., Taylor, K. A. and Thompson, D. T. (1968). *Reactions of Transition-Metal Complexes* (London: Elsevier).
Cardin, D. J., Cetinkaya, B. and Lappert, M. F. (1972). *Chem. Rev.* **72**, 545.
Casey, C. P. (1976). *Transition Metal Organometallics in Organic Synthesis*, vol. 1, p. 190. Ed. H. Alper (New York: Academic Press).
Chatt, J. (1951). *Chem. Rev.* **48**, 1.
Christensen, J. J. and Izatt, R. M. (1978), ed. *Synthetic Multidentate Macrocyclic Compounds* (New York: Academic Press).

Coates, G. E., Green, M. L. H. and Wade, K. (1968). *Organometallic Compounds*, vol. 2 (London: Methuen and Co. Ltd).
Cockerill, A. F., Davies, G. L. O., Harden, R. C. and Rackham, D. M. (1973). *Chem. Rev.* 73, 553.
Collman, J. P. (1965). *Angew. Chem. internat. Edit.* 4, 132.
Collman, J. P. (1975). *Accounts Chem. Res.* 8, 342.
Collman, J. P. (1977). *Accounts Chem. Res.* 10, 265.
Collman, J. P. and Roper, W. R. (1968). *Adv. Organomet. Chem.* 7, 54.
Corey, E. J. and Bailar, J. C. (1959). *J. Amer. Chem. Soc.* 81, 2620.
Cotton, F. A. and Lukehart, C. M. (1972). *Prog. Inorg. Chem.* 16, 487.
Cotton, F. A. and Wilkinson, G. (1972). *Advanced Inorganic Chemistry*, 3rd edit. (New York: Interscience Publishers).
Cowell, G. W. and Ledwith, A. (1970). *Quart. Rev.* 24, 119.
Davies, N. R. (1967). *Rev. Pure. App. Chem.* 17, 83.
Deeming, A. J. (1972). *M.T.P. International Review of Science, Inorganic Chemistry Series One*, vol. 9, p. 117. Ed. M. I. Tobe (London: Butterworths).
Dehand, J. and Pfeffer, M. (1976). *Coord. Chem. Rev.* 18, 327.
Dunitz, J. D. *et al.* (1973), ed. *Structure and Bonding*, vol. 16.
Dunlop, J. H. and Gillard, R. D. (1966). *Adv. Inorg. Chem. Radiochem.* 9, 185.
Eliel, E. L. (1962). *Stereochemistry of Carbon Compounds*, pp. 63 and 65 (New York and London, McGraw-Hill Book Co. Inc.).
Eschenmoser, A. (1970). *Quart. Rev.* 24, 366.
Falbe, J. (1970). *Carbon Monoxide in Organic Synthesis* (Berlin: Springer-Verlag).
Fischer, E. O. (1976). *Adv. Organometal. Chem.* 14, 1.
Freeman, H. C. (1967). *Adv. Protein Chem.* 22, 255.
Gaudemar, M. (1972). *Organomet. Chem. Rev.* 8A, 183.
Gillard, R. D. (1967). *Inorg. Chem. Acta. Rev.* 1, 69.
Gilman, H. and Morton, J. W. (1954). *Organic Reactions*, vol. 8, p. 258. Ed. R. Adams (New York: John Wiley and Sons).
Gokel, G. W. and Durst, H. D. (1976). *Synthesis*, p. 168.
Green, M. L. H. and Jones, D. J. (1965). *Adv. Inorg. Chem. Radiochem.* 7, 115.
Haines, R. J. and Leigh, G. J. (1975). *Chem. Soc. Rev.* 4, 155.
Halpern, J. (1970). *Accounts Chem. Res.* 3, 386.
Hammond, G. S., (1956). *Steric Effects in Organic Chemistry*, p. 425. Ed. M. S. Newman (New York: John Wiley and Sons).
Hartley, F. R. and Vezey, P. N. (1977). *Adv. Organometal. Chem.* 15, 189.
Heck, R. F. (1968). *Organic Syntheses via Metal Carbonyls*, p. 388. Ed. I. Wender and P. Pino (New York: John Wiley and Sons).
Hofer, O. (1976). *Topics in Stereochemistry*, 9, 111.
Huheey, J. E. (1972). *Inorganic Chemistry* (New York: Harper and Row).
James, B. R. (1973). *Homogeneous Hydrogenation* (New York: John Wiley and Sons).
Johnson, F. (1965). *Friedel Crafts and Related Reactions*, vol. 4, p. 45. Ed. G. A. Olah (New York: Interscience Publishers).
Jones, M. M. (1968). *Ligand Reactivity and Catalysis* (New York: Academic Press).
Jukes, A. E. (1974). *Adv. Organomet. Chem.* 12, 215.
King, R. B. (1964). *Adv. Organometallic Chem.* 2, 157.
King, R. B. (1970). *Accounts Chem. Res.* 3, 417.
Kitching, W. (1968). *Organomet. Chem. Rev.* 3A, 61.
Kotz, J. C. and Pedrotty, D. G. (1969). *Organomet. Chem. Rev.* 4A, 479.
Kroll, H. (1952). *J. Amer. Chem. Soc.* 74, 2036.
Kruck, Th. (1967). *Angew. Chem. internat. Edit.* 6, 53.
Leussing, D. L. (1976). *Metal Ions in Biological Systems*, vol. 5, p. 1. Ed. H. Sigel (New York and Basel: Marcel Dekker).
Lindoy, L. F. (1975). *Chem. Soc. Rev.* 4, 421.
McCleverty, J. A. (1968). *Prog. Inorg. Chem.* 10, 49.
McLendon, G. and Martell, A. E. (1976). *Coord. Chem. Rev.* 19, 1.
McQuillin, F. J. (1976). *Homogeneous Hydrogenation in Organic Chemistry* (Dordrecht: Reidel).

Maitlis, P. M. (1966). *Adv. Organomet. Chem.* **4**, 95.
Maitlis, P. M. (1971). *The Organic Chemistry of Palladium,* vol. 2, p. 82 (London: Academic Press).
Malatesta, L. and Bonati, F. (1969). *Isocyanide Complexes of Metals* (New York: John Wiley and Sons).
Mason, R. and Meek, D. W. (1978). *Angew. Chem. internat. Edit.* **17**, 183.
Mayo, B. C. (1973). *Chem. Soc. Rev.* **2**, 49.
Morrison, J. D., Masler, W. F. and Neuberg, M. K. (1976). *Adv. in Catal.* **25**, 81.
Morrison, J. D. and Mosher, H. S. (1971). *Asymmetric Organic Reactions* (New Jersey: Prentice-Hall).
Nelson, J. H. and Jonassen, H. B. (1971). *Coord. Chem. Rev.* **6**, 27.
Normant, J. F. (1972). *Synthesis,* p. 63.
Noyori, R. (1976). *Transition Metal Organometallics in Organic Synthesis,* vol. 1, p. 117. Ed. H. Alper (New York: Academic Press).
Olah, G. A. (1963), ed. *Friedel Crafts and Related Reactions,* vols. 1–4, (New York: Interscience Publishers).
Olah, G. A. (1973). *Angew. Chem. internat. Edit.* **12**, 173.
Orchin, M. (1966). *Adv. Catal.* **16**, 1.
Pearson, R. G. (1963). *J. Amer. Chem. Soc.* **85**, 3533.
Pettit, L. D. and Barnes, D. S. (1972). *Fortsch. Chem. Forschung,* **28**, 85.
Pizey, S. S. (1974). *Synthetic Reagents,* vol. 1, p. 295 (Chichester: Ellis Horwood).
Posner, G. H. (1972). *Organic Reactions,* **19**, 1.
Postgate, J. R. (1971), ed. *Chemistry and Biochemistry of Nitrogen* Fixation (London: Plenum Press).
Quinn, H. W. and Tsai, J. H. (1969). *Adv. Inorg. Chem. Radiochem.* **12**, 217.
Raphael, R. A. (1955). *Acetylenic Compounds in Organic Synthesis,* p. 40 (London: Butterworths).
Rathke, M. W. (1975). *Organic Reactions,* **22**, 423.
Rosenblum, M. (1974). *Accounts Chem. Res.* **7**, 122.
Rossotti, F. J. C. and Rossotti, H. (1961). *The Determination of Stability Constants* (New York: McGraw-Hill).
Satchell, D. P. N. (1977). *Chem. Soc. Rev.* **6**, 345.
Schrauzer, G. N. (1976). *Angew. Chem. internat. Edit.* **15**, 417.
Schriver, D. F. (1970). *Accounts Chem. Res.* **3**, 231.
Semmelhack, M. F. (1972). *Organic Reactions,* **19**, 115.
Seven, M. J. and Johnson, L. A. (1960), ed. *Metal-Binding in Medicine* (Philadelphia: Lippincott).
Seyferth, D. (1972). *Accounts Chem. Res.* **5**, 65.
Sillen, L. G. and Martell, A. E. (1964). *Stability Constants.* The Chemical Society, Special Publication no. 17.
Sillen, L. G. and Martell, A. E. (1971). *Stability Constants. Supplement no. 1.* The Chemical Society, Special Publication no. 25.
Snell, E. E. (1958). *Vitamins and Hormones,* **16**, 77.
Stern, E. W. (1968). *Catal. Rev.* **1**, 73.
Stille, J. K. and Lau, K. S. Y. (1977). *Accounts Chem. Res.* **10**, 434.
Storhoff, B. N. and Lewis, H. C. (1977). *Coord. Chem. Rev.* **23**, 1.
Taube, H. (1952). *Chem. Rev.* **50**, 69.
Thompson, D. T. and Whyman, R. (1971). *Transition Metals in Homogeneous Catalysis,* p. 147. Ed. G. N. Schrauzer (New York: Marcel Dekker).
Tolman, C. A. (1972). *Chem. Soc. Rev.* **1**, 337.
Traylor, T. G. (1969). *Accounts Chem. Res.* **2**, 152.
Treichel, P. M. (1973). *Adv. Organomet. Chem.* **11**, 21.
Trost, B. M. (1977). *Tetrahedron,* **33**, 2615.
Tsuji, J. (1969). *Accounts Chem. Res.* **2**, 144.
Verkade, J. G. (1972). *Coord. Chem. Rev.* **9**, 1.
Walton, R. A. (1965). *Quart. Rev.* **19**, 126.
Weiss, F. (1970). *Quart. Rev.* **24**, 278.

Wilds, A. L. (1944). *Organic Reactions*, vol. 2, p. 178 (London: Chapman and Hall).
Winterfeldt, E. (1969). *Chemistry of Acetylenes*, p. 267. Ed. H. G. Viehe (New York: Marcel Dekker).
Wojcicki, A. (1973). *Adv. Organomet. Chem.* 11, 88.
Woodward, R. B. *et al.* (1968). *J. Amer. Chem. Soc.* 90, 439.
Yamamoto, Y. and Yamazaki, H. (1972). *Coord. Chem. Rev.* 8, 225.

Index